OXFORD QUICK REFERENCE

A Dictionary of

Animal Behaviour

David McFarland graduated from Liverpool University in 1961 (Honours Zoology Class I). He then became a Domus Senior Scholar at Merton College Oxford, and a graduate student in Experimental Psychology, obtaining his D. Phil. in 1965. In 1964 he became Lecturer in Psychology at the University of Durham, and in 1966, Lecturer in Psychology at Oxford University and a Fellow of Balliol College. In 1974 he was appointed as University Reader in Animal Behaviour, and moved to the Department of Zoology Oxford. From 1978–79 he was Professor of Psychology at the University of Pennsylvania. He retired from Oxford in 2000 and took up a two-year appointment as Professor of Biological Robotics at the University of the West of England. He is now President of Casa Cantarilla, an Association of Teachers in the Arts and Sciences, located in Lanzarote, Spain.

The most authoritative and up-to-date reference books for both students and the general reader.

Accounting
Animal Behaviour
Archaeology
Architecture and Landscape Architecture
Art and Artists
Art Terms
Arthurian Literature and Legend
Astronomy
Battles
Bible
Biology
Biomedicine
British History
British Place-Names
Business and Management
Card Games
Chemical Engineering
Chemistry
Christian Art and Architecture
Christian Church
Classical Literature
Computing
Construction, Surveying, and Civil Engineering
Cosmology
Countries of the World
Critical Theory
Dance
Dentistry
Ecology
Economics
Education
English Etymology
English Grammar
English Idioms
English Literature
English Surnames
Environment and Conservation
Everyday Grammar
Film Studies
Finance and Banking
Foreign Words and Phrases
Forensic Science
Geography
Geology and Earth Sciences
Hinduism
Human Geography
Humorous Quotations
Irish History
Islam
Journalism
Kings and Queens of Britain
Law
Law Enforcement
Linguistics
Literary Terms
London Place-Names
Mathematics
Marketing
Mechanical Engineering
Media and Communication
Medical
Modern Poetry
Modern Slang
Music
Musical Terms
Nursing
Opera Characters
Philosophy
Physics
Plant Sciences
Plays
Pocket Fowler's Modern English Usage
Political Quotations
Politics
Popes
Proverbs
Psychology
Quotations
Quotations by Subject
Reference and Allusion
Rhyming
Rhyming Slang
Saints
Science
Scottish History
Shakespeare
Slang
Social Work and Social Care
Sociology
Statistics
Synonyms and Antonyms
Weather
Weights, Measures, and Units
Word Origins
Zoology

Many of these titles are also available online at www.Oxfordreference.com

A Dictionary of

Animal Behaviour

DAVID McFARLAND

OXFORD
UNIVERSITY PRESS

Great Clarendon Street, Oxford OX2 6DP

Oxford University Press is a department of the University of Oxford.
It furthers the University's objective of excellence in research, scholarship,
and education by publishing worldwide in

Oxford New York

Auckland Cape Town Dar es Salaam Hong Kong Karachi Kuala Lumpur
Madrid Melbourne Mexico City Nairobi New Delhi Shanghai Taipei Toronto

With offices in

Argentina Austria Brazil Chile Czech Republic France Greece
Guatemala Hungary Italy Japan Poland Portugal Singapore
South Korea Switzerland Thailand Turkey Ukraine Vietnam

Oxford is a registered trade mark of Oxford University Press
in the UK and in certain other countries

Published in the United States
by Oxford University Press Inc., New York

First edition published 2006

British Library Cataloguing in Publication Data
Data available

Library of Congress Cataloging in Publication Data
Data available

Typeset in Swift and Frutiger
by SPI Publisher Services, Pondicherry, India
Printed in Great Britain
on acid-free paper by
Clays Ltd, Bungay, Suffolk

ISBN 978-0-19-860721-2

Contents

Preface

This dictionary of animal behaviour aims to provide for students of biology and psychology who are interested in the behaviour of animals, and for naturalists, bird watchers, and animal enthusiasts who have similar interests. Most of the entries relate directly to animal behaviour, but the terminology of related subjects, especially ecology, physiology, and psychology, is also given some limited coverage.

The interdependence of much of the terminology used in the study of animals behaviour has necessitated extensive cross-referencing, and the reader may have to consult more than one entry to obtain a full understanding of a particular item. Some entries are illustrated by examples taken from nature, but these have been used sparingly, as a wealth of examples is readily available in the suggested further reading.

This dictionary is not intended to provide access to the behaviour of individual animal species, nor would this be possible in a book of this size. This book is designed to help those who wish to learn more about the study of animal behaviour as currently practised by professionals.

In order to provide a standard scientific classification of the animal kingdom that is accessible to the layman, the scientific names of animals used in this dictionary conform, as far as possible, with those used in *Grzimek's Animal Life Encyclopedia*.

The author is grateful to his commissioning editors, Michael Rodgers and Ruth Langley.

Abbreviations

AS	active sleep
dB	decibel
DNA	deoxyribonucleic acid
EEG	electroencephalograph
ESS	evolutionary stable strategy
h	hour
Hz	hertz
IQ	intelligence quotient
kHz	kilohertz
m	metre
m^2	squared metres
min	minute
ms	millisecond
nm	nanometre
QS	quiet sleep
REM	rapid eye (or ear) movements
s	second
sp. (pl. spp.)	species
SPL	sound pressure level
UCR	unconditional response
UCS	unconditional stimulus

abnormal behaviour Behaviour that occurs as a result of a pathological condition, including anxiety and *stress. Abnormal behaviour is sometimes seen in animals in captivity, where it may take the form of *stereotyped behaviour or *apathy.

acclimation A term used in laboratory experiments in which animals adapt to changes in a single environmental variable such as temperature. The term *acclimatization is reserved for the complex of adaptive processes that occurs under natural conditions.

acclimatization A form of reversible *adaptation by which an animal is able to alter its *tolerance of environmental factors. Under natural conditions acclimatization usually occurs in response to seasonal changes in climate. For example, temperature tolerance in fish is often directly correlated with monthly changes in their habitat temperature. Many fish also show temperature preferences that are related to their state of acclimatization.

action A form of *goal-directed behaviour, having the appearance of being *intentional. Normally applied only to humans.

action theory The body of philosophical theory that accounts for actions. There are many forms of action theory, and it is not possible to give a single, generally agreed upon, definition of an action. Probably, the most widely held view amongst philosophers is the causal theory, which maintains that the difference between a movement pattern and an action is that the latter is accompanied by a particular kind of mental event, which plays a causal role, whereas the former is not so accompanied.

Basically, action theories are goal-directed theories, because they envisage a mental state of affairs that is related to the goal (i.e. some form of knowing about the likely consequences) that is instrumental in guiding the behaviour.

active sleep That part of the *sleep cycle of birds and mammals during which there is a characteristic electroencephalograph (EEG) accompanied by rapid eye (or ear) movements (REM).

The sleep cycle is made up of alternating periods of quiet sleep and active sleep. The former is characterized by a synchronized EEG and no REM or signs of dreaming. The latter is characterized by desynchronized EEG (as in the awake animal), REM, and reports of *dreaming in humans, or signs of dreaming in other mammals. The periodicity of the sleep cycle is strongly correlated with the body size that is characteristic of the species. This may be due to the demands of *thermoregulation.

Thermoregulatory mechanisms function normally during quiet sleep, but give no response to thermal stress during active sleep. Such lack of thermoregulatory response is much more serious for small animals, whose body temperature is more easily influenced by ambient temperature than larger animals with a greater thermal capacity. Small animals have shorter periods of active sleep than large animals. It appears that each period of active sleep is terminated before it becomes thermally risky.

activity A measure of general activity employed in laboratory studies of animal behaviour. The measure is usually made mechanically, in a running wheel, stabilimeter (a cage with a tilting floor), or pedometer.

adaptation A term that has a number of meanings, all denoting some adjustment by the animal to changes in the environment. First, a distinction should be made between genotypic adaptation, in which the adjustment is *genetic and takes place through a process of *evolution, and phenotypic adaptation, in which the adjustment takes place within the individual animal on a non-genetic basis.

Phenotypic adaptation may be reversible or non-reversible, and may involve such processes as maturation, learning, and physiological adjustments such as *acclimatization. Irreversible adaptation may be observed at various points in the developmental history of an individual. Transient environmental changes may influence an embryo or larva so that a phenotypic characteristic of the animal becomes fixed. For example, in some bird species the young can become psychologically attached to members of another species, notably foster species. Similar processes apply to *food selection, *habitat selection, and *mate selection.

Another form of irreversible adaptation is *learning. Learning is essentially an irreversible process, for although animals may appear to forget or extinguish learned behaviour, the internal changes brought about by the learning process are permanent, and can be changed only by further learning.

Reversible adaptations enable the animal to adjust to local environmental changes, and usually involve rapid physiological or sensory changes. For example, changes in environmental temperature may induce adaptations in the form of *thermoregulation. Changes in light conditions may induce sensory adaptations in the mechanisms of *vision.

In summary, the term adaptation is used in a variety of ways, and biologists distinguish among: (1) evolutionary adaptation, which concerns the ways in which species adjust genetically to changed environmental conditions in the very long term; (2) physiological adaptation, which has to do with the physiological processes involved in adjustment by the individual to climatic changes, changes in food quality, etc.; (3) sensory adaptation, by which the sense organs adjust to changes in the strength of stimulation; and (4) adaptation by learning, which enables animals to adjust to a wide variety of environmental changes.

advertisement A form of *display usually shown by males holding a *territory. It serves both to attract females and ward off rival males. To be effective, such displays should be conspicuous and oriented towards likely recipients. The advertising type of bird *song, for example, should be easily locatable and should indicate the species, sex, and motivation of the singer.

aestivation A period of dormancy during the summer, in contrast to *hibernation, which is a period of winter dormancy. Aestivation enables animals to avoid extremes of climate, especially excessive heat or drought. It is most common in desert-living species, but is not confined to them.

aggregation A group of animals that forms as a result of individuals being independently attracted to a particular environmental feature, such as a source of food. Cases in which individuals are attracted towards each other, as in *flocking, are not considered to be aggregations.

Aggregation often occurs as a result of *habitat preference. For example, woodlice (*Porcellio* spp.) are usually found in moist places. These animals exhibit a particular form of *orientation, in which speed of locomotion is directly related to humidity. They run around more quickly in dry air than in moist, and consequently tend to stay longer in damp places.

aggression Damage, or threat of damage, to members of the same species. Similar behaviour to members of other species is usually classed as *predation or *defence. Thus a cat is not aggressive when catching a mouse any more than a swallow is aggressive when catching a fly.

Aggressive behaviour may involve actual attack, or *threat of attack. It usually involves posture and *display that is very typical of the species. For example, rattlesnakes (*Crotalus* spp.) do not bite each other when fighting, but each attempt to pin the other to the ground. Other forms of aggressive behaviour include bird *song, *scent marking, and *facial expression. These have all evolved, through a process of *ritualization, into specialized forms of *communication, meaningful only to members of the same species. There are sometimes cases of mistaken identity. Thus, an upright posture is a sign of aggression to a kangaroo, and a zookeeper should take care to adopt a stooped posture when entering a cage where there is a large male.

Aggression has evolved as a means of defending, or obtaining, resources such as food, *territory, or mates. An important theatre for aggression is male rivalry. In many species males compete for access to females, and this often results in aggressive behaviour, or even combat. Among red deer (*Cervus elaphus*), for example, The stags grow antlers each year, and in the autumn rutting season they challenge each other directly for ownership of females. Males challenge each other initially by roaring. These roaring matches enable *assessment of fighting potential, because a larger, fitter stag can usually roar faster. If the roaring contest is roughly equal, then the two stags may approach and walk parallel to each other, which enables the rivals to assess each other more closely, particularly with respect to body size. Many contests end at this stage, but some escalate into fighting proper. The stags interlock antlers and push against each other. The risk of permanent injury from such fighting is considerable. Usually the larger stag wins, but although winning may bring considerable benefits, these may be offset by the risk of loss of assets, and of future fighting potential. During fights, other males may attempt to steal the unguarded females. Moreover, to grow large antlers each year requires a large amount of energy and dietary materials, and successful stags are often exhausted by the end of the rutting season, and may die of starvation if there is a hard winter. This example serves to show that aggression is a complex matter, involving communication, bluff, assessment, and opportunism, as well as fighting ability.

agonistic behaviour The complex of *aggression, *threat, *appeasement, and *avoidance that often occurs during encounters between members of the same species. When encountering a stranger or rival, many animals exhibit a mixture of aggression and fear, which manifests itself in various forms of agonistic behaviour, including *ambivalent behaviour and *conflict. Agonistic behaviour is usually typical of the species, often taking the form of characteristic *displays.

alarm calls A vocal *alarm response, common amongst birds and mammals. They often include *auditory signals that make it difficult for a predator to locate the calling animal. To be difficult to locate, a sound should begin and end gradually, so that the predator cannot easily compare the times at which the sound reaches the two ears. For the same reason, it should be uniform in pitch. Low-pitched sounds are generally harder to locate than high-pitched sounds, and the alarm calls of many small birds are typically low-pitched, continuous calls, which begin and end gradually. In fact, these calls are so similar to each other that they often serve to warn members of more than one species. Some species, e.g. vervet monkeys (*Cercopithecus aethiops*) and chickens (*Gallus gallus domesticus*) have different alarm calls for different types of predator.

alarm responses Responses to signs of danger that act as a *warning to other animals. They normally take the form of specific visual, auditory, or olfactory warning signals. Not all responses to signs of danger should be considered as alarm responses. The house cricket (*Acheta domestica*), for example, simply stops chirping when it senses danger, and there is no evidence that this serves as a warning to others.

In those cases where an alarm response is likely to attract the attention of a predator, the individual that issues the warning may be disadvantaged. Such instances of apparent *altruism, where an individual endangers itself to the benefit of others, are thought to persist in evolution only when the other animals are close relatives of the animal issuing the warning signal.

In some cases, the nature of the warning signal is such that the alarm response incurs no disadvantage. For example, pigeons (*Columbia livia*) commonly feed in groups, and signal by means of flight *intention movements when they are about to fly away. A pigeon that signals its intention generally departs without disturbing the others. However, if a pigeon senses danger, it flies off without issuing any intention signal. The other pigeons then immediately take alarm and fly off also. The warning signal, which is the absence of intention movements, obviously does not disadvantage the initiator of the alarm, since it is the first to take to the air.

alertness A state of *arousal akin to *vigilance. The alert animal is in a state of readiness to detect any disturbance in the environment, whereas the vigilant animal is attuned to detect specific events, such as signs of a predator or prey.

allogrooming An aspect of social behaviour, in which one individual grooms another. In addition to removing parasites, such *grooming may serve to strengthen *social relationships, in which case it is often mutual. It may also serve an *appeasement function, reducing tension between rivals.

alpha male The top male in a dominance *hierarchy.

altricial young Young (usually birds) which hatch (or are born) at an early stage of development, and are helpless compared with *precocial young. Altricial young depend greatly on their parents for food and protection, while precocial young are less dependent.

If we compare altricial and precocial young at the same stage of physical development, such as the point which the feathers first appear, we find that the precocial bird is still inside the egg, while the altricial bird has already hatched. By the time the precocial bird hatches, the altricial juvenile has had the opportunity to learn a great deal, while the precocial bird has not yet begun to learn. For example, the altricial white-crowned sparrow (*Zonotrichia leucophrys*) learns the song of its species while a nestling. In contrast, chickens (*Gallus gallus domesticus*) are precocial, passing early infancy while still in the egg, but they have the innate ability to produce normal vocalizations even though never previously exposed to them.

altruism Self-destructive behaviour performed for the benefit of others. This biological usage contains no implications about intentions or motives. An important question is what is meant by benefit. One possibility is to regard the cost to the altruist and the benefit to the recipient as being measured in units of *inclusive fitness. This is defined in such a way that natural selection would not be expected to favour animals that improved the inclusive fitness of others at the expense of their own. Parental care would not qualify as altruism by this definition, for by caring for its young an animal increases its own fitness.

Another possibility is to define cost and benefit in terms of simple survival chances. An altruistic act then becomes one that decreases the altruist's chances of surviving, while increasing the survival chances of some other individual, the beneficiary. Parental care increases the survival chances of the offspring, while it decreases the life expectancy of the parent.

In genetic terms, genes for parental care tend to be preserved in the bodies of the surviving offspring, and such genes are, therefore, likely to increase in frequency relative to the genes that promote neglect towards offspring. Thus altruism at the individual level is a manifestation of 'selfishness' at the gene level.

Many young animals receive care from relatives other than their parents. Amongst Florida scrub jays (*Aphelocoma coerulescens*) about half the nests with young are attended by 'helpers', usually siblings of the new brood. Such helpers at the nest may contribute as much as 30% of the food that the nestlings consume. These acts of altruism at the individual level, are also 'selfish' at the gene level, because they benefit copies of helper genes in the bodies of relatives.

Amongst the social insects, such helping strategies have evolved much further. Most individuals in colonies of termites (Isoptera), and wasps (Tesebrantes), bees (Apoidea), and ants (Formicidae), are sterile female workers, who never reproduce themselves, but devote their lives to defending, feeding, and caring for their young siblings. The genes promoting such altruistic behaviour are passed on in the bodies of the males and the queen.

Altruism towards kin can be regarded as selfishness on the part of the genes responsible, because copies of these genes are likely to be present in relatives. Altruism could also be regarded as a form of gene selfishness if, by being altruistic, an individual could ensure that it was a recipient of altruism at a later date. The problem with the evolution of this kind of altruism is that individuals that cheated by receiving but never giving would be at an advantage.

It is possible that cheating could be countered if individuals were altruistic only toward other individuals that were likely to reciprocate. For example, when a female olive baboon (*Papio anubis*) comes into oestrus, a male forms a consort relationship with her. He follows her around, waiting for an opportunity to mate, and guards the female from the attentions of other males. However, a rival male may sometimes solicit the help of a third male in an attempt to gain access to the female. While the solicited male challenges the consort male to a fight, the rival male gains access to the female. Those males that most often give aid of this type, also most frequently receive aid. Thus, *reciprocal altruism can be demonstrated in this situation.

ambivalent behaviour Behaviour that indicates simultaneous tendencies to perform two incompatible activities. It is typical of *conflict situations. Sometimes the animal makes *intention movements towards both types of behaviour. Thus a half-tame moorhen (*Gallinula chloropus*), when offered food, may make incipient pecks towards the food while edging away from the outstretched hand.

Ambivalent behaviour is typical of *threat, and other situations in which there is motivational conflict. In many cases, it is thought to have evolved into a stereotyped *display, through a process of *ritualization.

ambush behaviour A form of *hunting by speculation. The predator usually lurks in a camouflaged fashion in a place that potential prey are likely to frequent. Only when the prey comes within range is an attack launched. For example, the mantid (*Parastagmatoptera unipunctata*) is a carnivorous insect that lies in wait throughout the day and catches flies that come within reach. It sits motionless among vegetation that resembles its own form and coloration. The prey is detected by the mantid's well-developed compound eyes, and is faced by an aiming movement of the head. The mantid then strikes the prey with a very rapid movement of the forelegs.

Some ambush predators place themselves so as to obtain a good view of potential prey. Thus grouper fish (*Mycteroperca* spp.) lurk low down in the water, and attack prey fish that swim above them, silhouetted against the light sky.

amensalism A type of interaction among animals in which one animal is affected adversely, while the other is unaffected. True amensalism probably does not occur in nature, but it may be approximated in cases of *competition, where one species is dominant over another, in the sense that one has priority in the use of resources. Absolute priority would imply that the dominant animal was unaffected by the competition.

anachoresis The habit of avoiding predators by living concealed in a hole, crevice, or some other retreat. Anachoretes must either emerge to feed and mate, or they must have specialized feeding habits, such as feeding from items suspended in a current. To mate they must have external fertilization, or be gregarious and mate amongst themselves.

An example is the polychaete worm (*Arenicola* sp.), which lives entirely in a burrow in the inter-tidal beach. It sucks water and food into the burrow, and has external fertilization. Many other marine species have a similar lifestyle.

Some animals are anachoretic during the day, and emerge at night to forage for food. Examples are to be found among scorpions (Scorpiones), millipedes (Myriapoda), centipedes (Chilopoda), voles (*Microtus*), rabbits (Leporidae), and badgers (Melinae).

animal economics The application of economic principles to animal behaviour. The animal is treated as a consumer, as in human microeconomics. It is assumed that the animal is a *rational decision-maker, maximizing *utility (or an equivalent) and spending time and energy in a way that parallels the expenditure of time and money by humans. The animal's *budget, or limit to expenditure, of energy or time, constrains its behaviour. Thus an animal cannot devote more energy to an activity than it has available at the time. Nor can it devote more time than is available (per day).

Experiments with animals show that they exhibit characteristic *demand functions, a concept borrowed from economics to express the relationship between the price and the consumption of a commodity. For example, if an animal expends a certain amount of energy on a particular activity, then it usually does less of that activity if the energy requirement is increased. The elasticity of demand functions in animals gives an indication of the relative importance of the various activities in the animal's repertoire, and demand functions are closely related to *behavioural resilience.

anthropomorphism The tendency to attribute human characteristics to animals. Human visitors to zoos frequently remark on the human-like behaviour of the animals they see there, and reports of animal behaviour frequently contain anthropomorphic assumptions.

Anthropomorphism has its counterpart in the behaviour of other animals. Cases of mistaken identity amongst animals generally result from the mis-recognition of particular features of the other species. For example, a cardinal (*Cardinalis cardinalis*) regularly fed a goldfish that learned to surface to obtain titbits. The cardinal was responding to the open-mouth stimulus, which is similar to that of its own young. The gape provides a powerful *sign stimulus, which is responded to even though other aspects of the situation are inappropriate.

The propensity of animals to respond to but a few of the many possible stimuli in a given situation is sometimes exploited by other species. For example, cuckoos (Cuculinae), which lay their eggs in the nests of other species, often produce eggs that are similar in size and colour to those of the host species. This *mimicry exploits the egg-recognition mechanisms of the host, which fails to make a discrimination that it is quite capable of making.

Similarly, humans are fully capable of distinguishing themselves from other species, but they do not always do so. They often have to be specially trained to resist the temptation to interpret the behaviour of other species in terms of their normal (human) behaviour-recognition mechanisms. In particular, humans have a tendency to interpret behaviour in terms of *intention, even when this is inappropriate.

anti-predator behaviour A form of *defensive behaviour, that affords protection against *predation. It may be passive, as with behaviour associated with *camouflage and *mimicry, or active, as in *alarm responses and *escape behaviour.

antithesis The principle that opposite emotional expressions, such as pleasure and anger, use opposing sets of muscles. This can occur either as part of *facial expression, or in the postures involved in *display.

anxiety *See* Stress.

apathy Listlessness, lack of responsiveness to stimuli.

aposematism *Advertisement of dangerous or unpleasant properties, that serves as a *warning to others.

appeasement Behaviour that serves to inhibit or reduce *aggression between members of the same species, in situations where *escape is impossible or disadvantageous. Appeasement often involves *displays in which weapons are hidden or turned away. For example, storks (*Ciconiidae*) greet their mates by placing their heads over their backs and clapping their bills. In this way they turn away their weapon and behave in exactly the opposite way to their normal *threat display.

Appeasement sometimes takes the form of ritualized juvenile behaviour, especially food begging. In gulls (*Laridae*), for example, food begging by the female seems to appease the male's aggression and often results in *courtship feeding.

appetite A combination of *hunger and the perceived quality of available food. Deprivation of food leads to hunger, which involves a complex of changes in the animal's internal state. In the absence of food, or any indicators of food availability, the hungry animal has no appetite. In the presence of cues to food availability, such as directly perceived food, or perception of stimuli normally associated with food, the animal develops an appetite, or *incentive to eat. The stimuli normally associated with food may include the place where the animal has obtained food in the past, the time of day that the animal normally feeds, the sight of other animals feeding, signs of prey, etc.

The extent of appetite is determined partly by hunger and partly by the perceived value of the food, including its palatability, calorie content, and accessibility. Appetite for particular constituents of the diet, such as minerals and vitamins, is common in animals. Such specific appetites have a strong influence on food selection. 'Cafeteria' experiments on animals, and on human infants, show that, when presented with dietary constituents in different containers, the subjects achieve a balanced diet over a number of days. Experiments have also shown a marked ability to select foods that meet nutritional deficiencies, and to avoid foods that contain poisons. Thus appetite is an important determinant of *food selection.

appetitive behaviour The active, *goal-seeking and exploratory phase of behaviour that precedes the more stereotyped *consummatory behaviour that the animal exhibits when it reaches its goal. Upon reaching the goal, appetitive behaviour normally ceases.

In some cases, there are difficulties in separating appetitive goal-seeking behaviour from other forms of behaviour. For example, the nest-building behaviour of the blackbird (*Turdus merula*) begins with the search for large twigs to form a foundation. Small twigs are then collected to form the sides. The nest cup is made from mud and lined with fine grass and hair. While we may wish to regard the completed nest as the bird's ultimate goal, there are difficulties in distinguishing between the appetitive and consummatory aspects of the nest-building behaviour. One possibility is to regard the search for each twig as an appetitive episode, and its placement in the nest as a consummatory response. Another possibility is to regard the whole chain of behaviour as appetitive, and completion of the nest as consummatory.

The concept of appetitive behaviour is not relevant to all types of behaviour. We would not normally think of animals as showing appetitive responses for *avoidance of noxious stimuli, or *alarm responses. Similarly, some appetitive behaviour, such as that shown by ambush predators that lie in wait for prey, is characterized by inactivity rather than by the active searching normally thought of as appetitive.

approach–avoidance conflict *See* Conflict.

arousal A state in which the animal is in a state of wakefulness, ranging from drowsy to very alert. It may be measured by responsiveness to standard stimuli, or by the electrical activity of the *brain. The degree of arousal is often accompanied by specific postures. For example, sleeping herring gulls (*Larus argentatus*) tuck the bill under one wing and open the eyes at intervals. When simply drowsy, they sit in a hunched posture, bill forward, opening the eyes more frequently. When resting, they sit with the head raised, and when fully alert, the neck is stretched, feathers sleeked, and the eyes fully open. This continuum of arousal behaviour has been shown to correlate strongly with responsiveness to standardized external stimulation.

The term arousal is also sometimes used in respect of *aggression and *sexual behaviour, where its meaning is similar, but these phenomena are more properly described in terms of *motivation.

assessment A prelude to *fighting during which the antagonists assess each other's fighting potential, by looking at weapons, body size, etc. Since such encounters can be won by bluff, there has often been an evolutionary trend towards exaggeration of weapons, etc. An example is the now extinct giant deer (*Megalocerus giganteus*) which had antlers measuring more than 3 m in span. These would have been very costly to produce (antlers are re-grown every year), and probably became too much of an energy burden as the climate became colder during the Ice Age. Modern deer assess rivals by means of roaring competitions. One stag usually retreats if the other can roar faster. A stag has to be in good condition to roar well, and this is probably a good indicator of fighting ability.

Similar ritualized contests are found in many species. Thus, some beetles engage in pushing matches, which the larger one usually wins, and male bighorn sheep (*Ovis canadensis*) charge each other and clash head on.

association A process involved in *learning, particularly in the type of situation initially studied by *Pavlov. When an animal responds to a particular stimulus it does so in the presence of other stimuli, such as marks on the ground, background sounds, or any of the normal features of everyday life. If one of these neutral stimuli is consistently paired with the stimulus that elicits the response, then it becomes associated with it through a process of *conditioning. Eventually, the previously neutral stimulus may come to elicit the response, even in the absence of the original stimulus.

In addition to its usage in describing learning behaviour, the term 'association' has theoretical implications. Learning by association implies that there is a linkage between stimuli, rather than a linkage between stimulus and response. Different theories of association may ascribe the linkage to stimulus substitution (one stimulus standing for another); contiguity of stimuli (contiguous stimuli are associated); or the learning of cause and effect relationships, in which an event (the cause) can cause another event (the effect) to happen, or not to happen. These different theories of association form part of the subject usually known as animal learning theory.

asymmetric contest An aspect of *evolutionary strategy in which the two participants in a contest do not have the same choice of strategies, or prospective payoffs. In contests between male and female, old and young, or the owner of a resource and a non-owner, an asymmetry may be perceived beforehand by the contestants, and will usually influence the choice of action. Such situations are usually analysed in terms of *game theory.

asymmetric games A type of *evolutionarily stable strategy, usually analysed in terms of *game theory. In a symmetric game, such as the hawk-dove game, the two contestants start in identical situations, and have the same choice of strategies and the same prospective payoffs. There may be a difference in strength or size between them, but if this is not known to the contestants, it cannot affect their choice of strategies.

In an asymmetric game there is usually a perceived difference between the contestants which affects their choice of strategies. The contestants may differ in sex, size, age, or a resource such as territory. For example, an individual could behave like a hawk when it is the owner of a territory and like a dove when it is an intruder into the territory of another. That is, when holding a territory, it fights to kill or injure the opponent, even though there is a risk of injury to itself. When an intruder, it threatens its opponent but avoids serious fighting. This is called a bourgeois strategy.

We now assign fitness-increment payoffs to the consequences of encounters involving hawks and doves. Let the winner of the contest score +50 and the loser, zero. Let the cost of wasting time in a display be −10 and the cost of injury be −100. When a bourgeois meets either hawk or dove, we assume it is an owner half the time and therefore plays hawk, and an intruder half the time and therefore plays dove. Its payoffs are therefore the average of hawk and dove. When bourgeois meets bourgeois, on half the occasions it is an owner and wins, while on half the occasions it is an intruder and retreats. There is never any cost of display or injury. When two bourgeois meet the average payoff is +25, more than could be gained by invading doves or hawks, who would get +7.5 and +12.5 respectively.

	Opponent		
Attacker	Hawk	Dove	Bourgeois
Hawk	½ (50) + ½ (−100) = −25	+50	+12.5
Dove	Zero	½ (50 − 10) + (−10) = +15	+7.5
Bourgeois	−12.5	+32.5	+25

The formulae indicate the average payoffs to the attacker.

The bourgeois strategy is an *evolutionarily stable strategy. Asymmetry of ownership is used as a convention to settle contests even when ownership alters neither the payoffs nor success in fighting. The same is true for any other asymmetry, provided it is unambiguously perceived by both contestants.

attack Assault upon prey or rival, using weapons designed to kill or maim. In the context of predatory *behaviour, attacks are usually the culmination of a period of *ambush or *hunting.

attention A perceptual process influencing selective response to stimuli. When in the presence of many relevant stimuli, animals usually respond selectively to one or a few of them. One of the processes determining this selectiveness is attention.

What is attended to may be influenced by the salience of the stimulus, the *motivation of the animal, and previous *learning by the animal. The phenomenon of selective attention is closely related to the concept of *search image. For example, carrion crows (*Corvus corone*) have been shown to respond selectively to a particular type of prey, even when other equally likely types are available. The birds were trained to search among beach stones for pieces of meat that had been hidden under mussel shells painted various colours. The birds would concentrate on shells of one colour, ignoring others. If a bird turned over a shell of a particular colour and found no meat, then it would usually start searching for shells of another colour. Similar phenomena have been demonstrated in laboratory experiments on a number of species of fish, bird, and mammal.

For example, if pigeons (*Columbia livia*) are trained to discriminate among stimuli of different colours, or among stimuli of different shapes, and they are then required to learn a new *discrimination in which the stimuli differ in both colour and shape, they show transfer of learning. Those that previously learned a colour discrimination learn the new discrimination more quickly than those that previously learned a shape discrimination, if colour is the relevant (i.e. rewarded) cue. Whereas those that previously learned a shape discrimination do better if shape is the relevant cue in the second task. In other words, in the first task the animals learn to discriminate among particular stimuli, but they also learn that they must pay attention to the colour (or shape) of the stimuli. This prior learning may benefit them in the second task.

auditory senses *See* Hearing.

auditory signals An aspect of *communication in which certain properties of sound have been exploited during evolution. Sound production can be turned on and off rapidly, and varied in other ways, so as to make possible an immense variety of signals. These can vary in pitch (frequency of vibration), loudness (although this may be confounded by distance), and temporal pattern.

A limit to the effective distance of auditory signals is the loudness that the signaller can achieve. This is directly related to the size of the sound-producing organ, and therefore of the animal. The mole cricket (*Gryllolalpa vineae*) increases its acoustically effective size by digging a double conical horn-shaped burrow, in which it sings. The burrow acts as a megaphone.

In addition, various physical properties of the environment are relevant to the distance that auditory signals travel. *See* Vocalization.

Sound localization depends partly upon the *hearing apparatus of the animal, and partly upon the nature of the auditory signal. A series of short staccato notes, each including a wide spectrum of frequencies provides an easily locatable sound. This is characteristic of the chink-chink *mobbing call of many species of songbird. The mobbing call summons other birds to help in harassing a ground-based, or perched, predator. Sounds that fade in and fade out gradually are difficult to locate. Such signals are characteristic of the *alarm calls of small birds.

automimicry A form of *mimicry in which, among members of the same species, some are palatable to predators and some are distasteful. Predators that attempt to eat the unpalatable ones quickly learn to avoid them. Since they look identical, the palatable ones are also avoided. Thus they gain protection by looking like unpalatable individuals, without carrying the unpalatable substances.

autonomic A division of the vertebrate nervous system serving internal organs, such as the heart, blood vessels, lungs, intestines, and also certain glands. It is controlled by the *brain and supplies the internal organs with two types of nerve supply that have antagonistic effects. The **sympathetic** nerve pathways have an emergency function, and become active under conditions of exertion or *stress. They have the effect of accelerating heart rate, dilating air passages to the lungs, increasing the blood supply to the muscles, and reducing intestinal activity. The **parasympathetic** pathways serve a recuperative function, restoring the blood supply to normal and generally counteracting the effects of sympathetic activity.

The autonomic nervous system influences behaviour in a number of important ways. It is responsible for many of the external signs of *fear, such as feather sleeking in birds, and of *aggression, such as hair erection in dogs.

autoshaping A type of learning in which the animal develops as an *operant without special training or *shaping. In pigeons (*Columbia livia*), key pecking is typical of *operant conditioning experiments. Usually, the key is illuminated and the light is extinguished when the key is pressed. If pigeons are presented with repeated pairing of key illumination and food reward, they begin to peck the key without any prior training.

This autoshaping phenomenon is thought to be the result of a straightforward *classical conditioning situation. The lighted key is paired with grain, and the pigeon comes to peck the key as if it were grain. The point can be demonstrated even more clearly if key illumination is paired with a reward to which pecking is not the pigeon's natural response. This can be done by offering a sexual reward. Paired male and female pigeons are housed in adjacent chambers, separated by a sliding door. Once a day the stimulus light is turned on

a

and the sliding door is removed, so that the male can begin his courtship display for a minute or so. Within five to ten trials the male pigeon begins to make conditioned sexual responses towards the stimulus light. The pigeons direct their courtship towards the light, thus behaving as if it were a female.

autotomy The ability to constrict, and break off, part of the body when attacked by a predator. Many lizards (*Draco* spp.) have tails that are more brightly coloured than the rest of the body, and this serves to divert an attack towards the tail. After detachment, the tail continues to writhe for some time, thus diverting the predator from the escaping lizard. The tail regenerates within a few weeks.

Autotomy is also found in marine molluscs (*Mollusca*) and polychaete worms (*Arenicola* sp.).

aversion Antipathy to food, which may be *innate or learned. Many animals have an inborn aversion to poisonous plants or animals, and many are also capable of rapid *aversive learning.

aversive learning A rapid form of *conditioning to the sight, smell, or taste of (usually novel) foods, which occurs as a result of sickness. If an animal consumes a substance with a distinctive smell or flavour, and is later subjected to toxic after-effects produced by such independent means as X-irradiation or injection of poison, it will avoid consuming the distinctive substance in the future. The animal behaves as if it thinks that consumption of the substance made it sick. This specific aversion will not occur if consumption occurs without being followed by toxicosis, or if toxicosis occurs in the absence of previous consumption. It differs from the types of learning usually investigated in that it can occur after a single pairing, even when the interval between ingestion and toxicosis is a number of hours.

avoidance Behaviour that is a form of defence against circumstances that might be harmful. Physically avoiding places where predators may lurk, and choosing not to consume a potentially poisonous substance, are typical examples. Unlearned avoidance reactions are part of the animal's *innate make-up. For example, many young birds give distress calls and attempt to hide, when a silhouette of a hawk (Accipitridae) passes overhead. No experience with predators is required for this to happen.

While innate avoidance behaviour is widespread amongst animals, the ability to learn new avoidance behaviour is important as well.

According to one viewpoint, the acquisition of learned avoidance is accomplished in two stages. First, the animal learns to *fear certain stimuli in the environment, as a result of *association with an aversive outcome. For example, consumption of a new food may be followed by sickness. Secondly, the animal learns behaviour that reduces the fear. This may take the form of passive avoidance, such as not eating novel foods, or active avoidance, such as *escape from the presence of a strange animal.

Much avoidance behaviour is neither wholly innate nor wholly learned. Many animals have species-specific *defence reactions. For example, some may struggle to escape when caught by a predator, while others feign *death. New avoidance behaviour is more quickly learned if it is similar to the species-specific defence behaviour of the animal concerned. For example, mice are naturally thigmotaxic (stay close to a wall) when frightened. Mice (*Murinae*) quickly learn to jump on to a box next to a wall to avoid a noxious stimulus, but have difficulty learning to jump on to a box in an open space.

avoidance learning The process of *learning to avoid situations that appear to be dangerous. Natural stimuli that induce innate *avoidance behaviour include those associated with predators. In its natural habitat, an animal that learned how to escape from predators by means of a trial-and-error procedure would survive for a few trials only. Animals usually have innate defence reactions that are modifiable by learning, and do not learn this kind of behaviour from scratch. Therefore, much depends upon the response that the animal is required to make to learn successful avoidance. Thus, pigeons (*Columbia livia*) readily learn to press a treadle to avoid shock, but have difficulty learning to peck a key. Rats (*Rattus norvegicus*) learn to run when escape is possible, and to freeze when it is not possible.

bait shyness A form of *avoidance learning in which the animal quickly develops an *aversion towards a novel food or bait. Attempts to eradicate local rat populations by using poison have had limited success, as a result of the rat's proverbial bait shyness. Rats (*Rattus norvegicus*) and mice (*Murinae*) tend to avoid unfamiliar foods and to sample them tentatively. A rat that takes a small amount of poisoned food and survives will never touch the same type of food again. This rapid aversive learning is found in many mammals and birds.

balance A type of mechanical *sense which enables the animal to maintain a stable position with respect to gravity. This requires a sense of equilibrium, which is provided by specialized mechanoreceptors in many species.

In vertebrates, these mechanoreceptors are found in the inner ear, separate from the hearing organ proper. The tips of hairs in the sensory cells are embedded in a gelatinous concentration of calcareous particles, the statoliths. When the head is tilted, the weight of these statoliths shears the hairs sideways, stimulating the receptor cells.

Many invertebrates have similar arrangements Octopus (*Octopus* spp.) for example, has a pair of statocysts, information from which controls the position of the eyes. Crustaceans have statoliths (sand particles contained within the statocysts) which are lost during each moult. The animals use their claws to put grains of sand into the statocysts at the base of the antennula. If forced to use iron filings instead of sand, they behave towards a strong magnet as they normally do towards gravity.

Batesian mimicry A form of *mimicry in which a predator that avoids a noxious animal producing a particular signal (the model) is deceived into avoiding an edible mimic that produces a similar signal. It is of advantage to the mimic, but of no advantage to the model.

The model is usually an animal with a nasty taste, a sting, or spines, but plants and inanimate objects can also be mimicked. For example, stick insects (Phasmida) closely resemble twigs that are inedible to predators.

behavioural ecology A branch of *evolutionary biology, in which the issue is the way in which behaviour contributes to survival and reproduction, and is dependent upon *ecology. The evolutionary persistence of a trait, such as a particular aspect of behaviour, depends upon its contribution to the survival and reproduction of the individual carrying the trait. Therefore, when we see an animal behaving in a particular way, we can ask how the behaviour contributes to survival and reproduction under the ecological circumstances.

behavioural final common path A theoretical concept, based upon the idea that there are activities for which the animal may have a strong *tendency, but which cannot be performed simultaneously. The behavioural final common path is a (hypothetical) mechanism by which the animal decides among incompatible tendencies. It is assumed that the strongest tendency wins, and is expressed as overt behaviour, while behavioural expression of all other potential activities (i.e. tendencies) is inhibited.

behavioural genetics The study of the genetic inheritance of behavioural traits. Methods of study include selective breeding experiments, *heritability studies, and DNA analysis.

behavioural resilience A measure of the extent to which an activity can be squashed in terms of time by other activities in the animal's repertoire. It also reflects the importance of an activity in the long-term sense. Resilience is mathematically related to *demand functions.

behavioural sampling Occasional selection of a less-preferred item in a *choice situation. This is thought to be a strategy by which some animals keep check of alternatives, even though they gain less by selecting that alternative.

behaviourism A school of psychology, launched by J.B. Watson (1913), which was very influential during the first half of the 20th century. The behaviourists would allow only observable stimuli, muscular movements, and glandular secretions to enter into explanations of behaviour. In order to account for complex behaviour in stimulus–response terms, they postulated covert or implicit stimulus–response relationships. The unobservable processes mediating stimulus and response were said to be comprised of incipient movements and movement-produced stimuli.

binocular vision The ability to focus on an object with both eyes simultaneously. Binocular vision is restricted to vertebrates, which have evolved an eye with special features aiding the perception of depth and distance. An important cue to distance is the extent of convergence of the two lines of sight. These are adjusted, via the extraocular muscles of the eyes, until the two one-eyed images are unified into a single object. Focusing on closer objects requires more convergence than focusing on objects further away.

Another cue to distance is retinal disparity. When two spatially separated eyes look at an object, each retina receives a slightly different view of the object. The disparity between these two images varies with the distance between the eyes and the object. The closer the object, the more disparate are the images.

Vertebrate species do not possess binocular vision to the same extent. For example, primates have extensive binocular fields of vision, whereas in fish and birds only part of the visual field is viewed by both eyes, the rest being monocular. Some species, including some whales (Cetacea), penguins (Spheniscidae), and fish, have no binocular vision. Such animals have a panoramic visual field, enabling them to detect predators, or prey, more easily. Thus there is a trade-off between the advantages and disadvantages of binocular vision, which is tailored to the lifestyle of the species.

biological clock A mechanism, internal to the animal, that has a rhythmic influence upon its physiology and behaviour, synchronizing them to cyclic changes in the environment. This is done in three main ways: (1) there may be a direct response to various changes in external (exogenous) geophysical stimuli; (2) there may be an internal (endogenous) rhythm that programs the animal's behaviour in synchrony with the exogenous temporal period, particularly a 24-hour or a 365-day period; or (3) the synchronization mechanism may be a combination of (1) and (2).

An animal may use many features of the external environment to gain information about the passage of time. The most important of these is the apparent movement of celestial bodies, such as the sun, moon, and stars. Such influences have been much studied in birds and in bees (Apidae). In addition, it is possible that animals can obtain time cues from changes in environmental temperature, barometric pressure, and magnetic phenomena.

Endogenous daily *rhythms are termed *'circadian' and usually fall short of a 24-hour periodicity. Annual endogenous rhythms, termed *'circannual', are usually less than 365 days. Since endogenous rhythms tend to deviate gradually from that of the exogenous cycle, an organism must have a means of synchronizing its endogenous rhythms with the cycles of external events. The term '*zeitgeber*' (meaning time-giver) is used for the external agent that entrains the endogenous rhythm to the external environment. For example, a 24-hour cycle of external temperature, with an amplitude of 0.6°C, is sufficient to act as a *zeitgeber, and to entrain the activity cycles of lizards.

biome A characteristic *ecosystem. Plant communities provide a variety of possible *habitats, that animals are able to exploit. The association of plants and animals, together with the physical features of the habitat, constitute an ecosystem. About ten general types of terrestrial ecosystem are found in the world. These are known as biomes. They are arctic tundra, northern coniferous forest, temperate forest, temperate grassland, chaparral (mediterranean vegetation), tropical rain forest, tropical savannah grassland, and desert.

boredom A supposed mental state that results from excessive *leisure, repetitive stimulation, or monotonous circumstances. Boredom is assumed to be an aversive state that an animal will attempt to avoid. Excessive boredom may be associated with stereotyped behaviour and *stress.

brain Part of the central, as opposed to the peripheral, nervous system of an animal. In invertebrates the central nervous system is located ventrally, whereas in vertebrates it is dorsal.

Many invertebrates have dual ventral nerve cords with anterior ganglia, that may be fused into a single brain. The size of the brain varies greatly with species. The octopus (*Octopus* sp.) brain contains about 170 million nerve cells, whereas the brain of a large crustacean contains about 100 000 nerve cells.

The organization of the vertebrate nervous system is distinct from that of invertebrates, although it is not always more complex. The pattern of embryological development of the brain is remarkably constant throughout the vertebrates. A dorsal neural tube develops into three distinguishable parts: the forebrain, midbrain, and hindbrain. These three hollow vesicles are traditionally associated with the olfactory, visual, and auditory senses, respectively.

The vertebrate central nervous system comprises the brain and spinal cord, which are enclosed by the bones of the skull and vertebral column. Vertebrate brains vary considerably in size, but larger animals tend to have larger brains because of the greater number of sensory fibres entering the brain, and the greater number of motor fibres leaving the brain to control the muscles. Taking into account differences in body size, the relative brain size of fish, amphibians, and reptiles is constant. The brains of birds and mammals are larger. The biggest brains are found among the primates and sea mammals. The human brain is three times as large as would be expected for a non-human primate of the same body size.

breathing Strictly, the movement of air in and out of lungs, as part of the function of gas exchange. However, many animals have no lungs, perform gas exchange in other ways, and the term 'breathing' is often applied to them. Respiration involves uptake of oxygen and the production of waste carbon dioxide.

The cells must be provided with a continuous supply of oxygen and the waste gases must be removed. In insects and spiders (*Arachnida*), air is presented directly to the cells via a system of tubes (tracheae), but in most other animals a circulatory system carries oxygen in the blood from a part of the body exposed to the air. Carbon dioxide travels in the opposite direction, from the cells to the external environment. The part of the body exposed to the environment, through which gas exchange takes place, may be the general body surface, gills, lungs, or tracheae.

breeding cycle A cycle of sexual and parental behaviour and physiology. Among animals living at middle and high latitudes, the breeding cycle is usually seasonal.

The reproductive physiology of seasonal breeders is geared to the annual cycle of environmental change in such a way that peak abundance of food, or adverse environmental conditions, are anticipated. First, changes in ambient temperature, day length, or other environmental cues provide stimuli that induce physiological change at a particular time of year. Secondly, the physiological changes are programmed on a seasonal basis by means of an endogenous *circannual rhythm.

budget The limit to expenditure, of energy or time, that constrains animal behaviour. An animal cannot devote more energy to an activity than it has available at the time. Nor can an animal devote more time than is available (per day). When an animal has a choice between activity A and activity B, the consequences of both these activities entail the expenditure of energy and time. This limit on energy (or time) means that whatever is spent on A cannot be spent on B. When time spent on A or B (e.g. eating and drinking) is plotted in terms of the normal consequences of these activities (i.e. food and water obtained), we can see (Figure 1) that the more time or energy spent on A, the less can be spent on B. The sum of the energy (or time) that can be spent is expressed as a budget line on the graph.

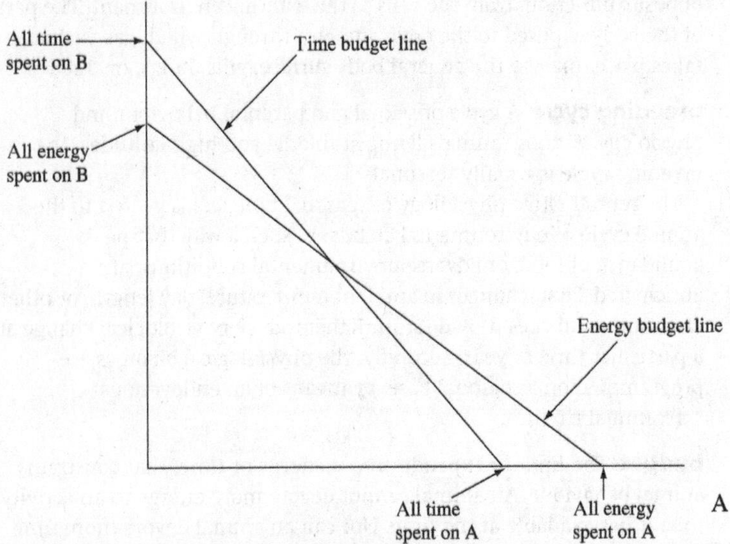

Fig. 1. Consequences of doing A plotted against consequences of doing B. Possible combinations of A and B must lie below both budget lines.

C

camouflage A form of visual deception, by means of which an animal can elude predators, or a predator may lurk undetected, awaiting prey.

An object can be detected easily against a particular background if it has a sharp outline, casts strong shadows, or is a different colour from the background. Most techniques of camouflage are designed to minimize these differences. An animal's body outline can be concealed by disruptive coloration, or by morphological projections, as in the frogfish (*Histrio histrio*), which lives among Sargassum seaweed (*see* Figure 2). The coloration of many species of frog and bird includes eye stripes that disrupt the otherwise conspicuous circular eye. Revealing shadows may be suppressed by counter-shading, in which the shadows normally cast on the underside of the body are counteracted by lighter coloration of those parts.

For camouflage to be effective, the animal must remain motionless. Thus there will be times when this *cryptic behaviour is incompatible with other essential activities. Some species remain motionless during the day and become active at night, or at dawn and dusk. Others,

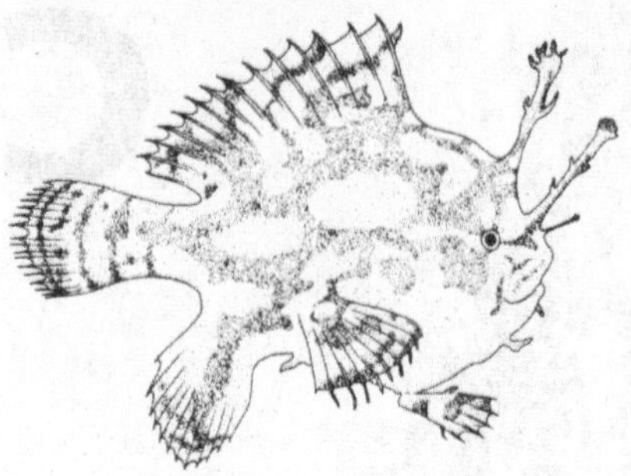

Fig. 2. The frogfish (*Histrio histrio*).

including many butterflies and moths, are active during the day, and take up a cryptic position as an *alarm response. Many studies have demonstrated the importance of appropriate background for animals depending upon camouflage. The dark-coloured moth (*Catocala antinympha*), given a choice of black or white backgrounds, usually rests on the black background, while the pale butterfly (*Campaea perlata*) chooses the white background. When these moths are painted to render them conspicuous against their chosen background, their behaviour does not change.

cannibalism The eating of members of one's own species, alive or dead. Cannibalism can take two forms, active and passive. The first refers to animals that actively hunt to kill and eat members of their own species. This may include killing and eating offspring (prolicide), and killing and eating siblings (fratricide). Both these occur in a variety of species.

Passive cannibalism is the eating of members of one's own species once they are dead. This is a normal aspect of *feeding behaviour in many species. The main factors that are known to influence cannibalism in animals are overcrowding, stress, and sexual rivalry.

central nervous system That part of the nervous system comprising the brain of invertebrates, or the *brain and spinal chord of vertebrates. To be contrasted with the *peripheral nervous system.

chain responses Behaviour sequences in which each movement or activity brings the animal into a situation where the next is evoked. Examples at the reflex level include walking and chewing, in which one component of the behaviour, such as stepping out, or opening the mouth, stimulates the next component through the stretching of muscles or pressure on certain parts of the limb.

Chain responses involving the behaviour of the whole animal generally depend upon one activity bringing the animal into a new external situation that leads to the next activity. Cases where the next activity is a direct result of changes in the environment caused by the previous activity, are sometimes known as *stigmergy.

Chain responses of a less rigid nature may occur in *social interactions, where the behaviour of one participant stimulates a response from another, and this stimulates a change in the behaviour of the first participant.

character displacement The tendency of closely related species to diverge in characteristics that reduce *competition between them. Darwin's finches (*Geospizini*), which inhabit the Galapagos Islands, are a well-studied example of this phenomenon.

cheating An *evolutionary strategy, in which some members of a species gain an advantage over others by not conforming to the social norm. For example, *alarm signals incur a cost in attracting the attention of a predator to the individual giving the alarm. All members of a group benefit from the alarms given by colleagues, but the situation is open to cheating. It might pay an individual to save time and reduce risks by reducing the time it spends being *vigilant in a group situation. Some species have mechanisms that reduce this possibility. Pigeons (*Columbia livia*) feeding in a flock give a special *intention signal when they are about to depart. If they depart in a hurry, without giving this signal, then all the other birds fly off in alarm. Thus the vigilant pigeon, by flying away without giving the 'goodbye' signal, is warning the other members of the flock, yet it gains an advantage by flying away first. This type of cheating is assumed to be genetically based, and it occurs in a wide variety of situations, including *courtship, *feeding, and *fighting.

chemoreception The capability of detecting chemical substances and of monitoring their concentration. The distinction between the *gustation and *olfaction cannot be made in many animals. Chemoreceptors occur both externally and internally. Strictly, every nerve cell acts as a chemoreceptor in that it reacts specifically to substances released by other nerve cells, but the term is usually reserved for those specialized *sensory cells that are designed to detect chemical substances in the environment (exteroceptors) and in the body fluids (interoceptors).

The mechanisms of chemoreception involve the recognition of specific molecules by receptor sites on cell membranes. Whether this recognition occurs on the basis of chemical action, molecular shape, or both, is not fully understood. Thus we do not know what sugar and saccharine have in common that makes them both taste sweet to blowflies, rats, monkeys, and humans. Both taste and smell depend upon chemoreceptors. In the traditional sense, smell is concerned with the detection of low concentrations of airborne substances, while taste results from direct contact with relatively high concentrations of chemical substances. In both cases, however, the chemicals are presented to the receptor in solution, and the distinction is difficult to justify in some animals, especially those that live in water. Nevertheless, in many animals there is a neurological distinction, in that some nerves are concerned with relaying olfactory messages, and others gustatory messages. In the blowfly (*Phormia regina*), for example, chemoreceptors in the antennae detect small quantities of airborne substances, and chemoreceptors in the tarsi (feet) are capable of detecting salt, sugar, and pure water. In vertebrates, the sense of taste is relayed via the facial (VII) and glossopharyngeal (IX) cranial nerves, while the sense of smell is transmitted by the olfactory nerve (I).

choice A situation in which the animal can make a simultaneous *discrimination amongst alternatives. Normally, an animal will choose the alternative that brings it the greatest short-term gain. Sometimes, however, animals show *behavioural sampling of lesser alternatives. This is thought to be a strategy that results in a long-term gain in *utility.

circadian rhythms Endogenous *rhythms in physiology and behaviour, that usually have (in isolation) slightly less than a 24-hour periodicity. The endogenous rhythm is due to an internal *biological clock.

circannual rhythms Endogenous *rhythms in physiology and behaviour, that usually have (in isolation) a periodicity slightly less than 365 days. The endogenous rhythm is due to an internal *biological clock.

classical conditioning The type of *conditioning first studied by *Pavlov, in which the animal forms an *association between a significant stimulus (the unconditional stimulus) and a neutral stimulus (the conditional stimulus). After repeated exposure to the paired stimuli, the animal's response to the unconditional stimulus (the unconditional response) can be elicited by the conditional stimulus alone. This response (the conditional response) is not always exactly the same as the unconditional response.

clock *See* Biological clock.

coadaptation Mutual adaptation: separate structures, or facets of behaviour are designed (by natural selection) specifically for interaction with each other. Coadaptation may occur among parts of a single organism. For example, certain light-coloured moths prefer to rest on light-coloured surfaces, and dark-coloured moths on dark surfaces. This coadaptation of coloration and behaviour functions to improve *camouflage.

Coadaptation between organisms may be intraspecific or interspecific. The former refers to coadaptation between organisms of the same species. For example, in *communication it is important that a signal sent by one individual is coadapted to the receiving apparatus of another.

Interspecific coadaptation refers to complementary adaptation between members of different species. Thus pollinating insects possess morphology and behaviour that is tailored specifically to the requirements of certain plants. Likewise, the plant structure is coadapted to specific pollinating insects. Other examples can be found in cases of *symbiosis between different animal species.

coefficient of genetic relatedness, *r* A measure of the probability that a gene in one individual will be identical by descent to a gene in a particular relative. An equivalent and alternative measure is the proportion on an individual's genome that is identical by descent to the relative's genome. Note that the genes must be identical by descent and not simply genes shared by the population as a whole. This means that r is the probability that a gene common to two individuals is descended from the same ancestral gene in a recent common relative.

The coefficient of genetic relatedness is important in understanding the concepts of *inclusive fitness and *altruism.

cognition Mental processes that are presumed to occur within the animal, but which cannot be observed directly. In its stricter sense, cognition refers to a particular kind of *knowledge: namely 'knowing that' rather than 'knowing how'. Cognition is the manipulation of explicit knowledge.

In its more general sense, cognition refers to any kind of mental abstraction of which an animal seems to be capable. For example, field studies show that chimpanzees (*Pan troglodytes*) sometimes take a roundabout route to a destination, suggesting that they have a mental picture of the spatial relationships of objects within the environment. In the study of *navigation, *problem solving, *social interactions, *deceit, *language, and *thinking in animals, scientists have found it necessary to postulate cognitive processes. Such suggestions have proved to be controversial, and the question of whether animals can think remains an open question.

coloration The colour patterns on an animal's body surface. These are usually tailored to the animal's lifestyle, and have important functions in relation to *advertisement, *camouflage, *mimicry, *thermoregulation, and *warning.

In the course of evolution, animals develop colour patterns appropriate to their *niche, including the visual capabilities of other species in the community. Thus, nocturnal animals are not usually brightly coloured, because the mechanisms of *colour vision do not work well in conditions of low illumination. There is often some competition between different pressures of *natural selection, usually between the advantages of camouflage and those of conspicuousness. Some animals, such as grasshoppers (*Trilophidia* sp.), attempt to obtain the best of both worlds by remaining camouflaged when motionless and displaying bright colours when moving. This is called **flash coloration**.

colour vision The ability to discriminate among light sources of differing wavelengths. The wavelength of light is a physical property, and the colour of light is a subjective property that depends upon the characteristics of the observer's visual and nervous systems. These vary greatly from species to species, and it is impossible to find out the apparent colour that a given object may have for an animal other than humans. Animal behaviour studies are more concerned with the extent to which animals have the ability to distinguish and recognize different lights through their wavelength composition.

In order to achieve colour vision, it is necessary for the animal to have neural receptors that are differentially sensitive to wavelength. In most cases this is achieved by the receptors having photosensitive pigments that absorb light more readily at some wavelengths than at others. In some cases (e.g. some lizards and birds), colour differentiation is achieved by means of coloured oil droplets that act as filters, and alter the light reaching the receptor.

*Vision that depends upon only one type of photopigment is **monochromatic**. Such is the case during night vision in humans. However, if an animal has two separate types of receptor, with different spectral sensitivities, colour vision becomes possible. Such an animal is the grey squirrel (*Sciurus carolinensis*). Because the spectral sensitivity curves overlap, each wavelength of the spectrum will result in a specific ratio of the outputs of the two receptor types, and these will be largely unaffected by changes in the intensity of light. So, provided the squirrel's brain has some mechanism for detecting these ratios, **dichromatic** colour vision, independent of changes of light intensity, is achieved.

The squirrel's colour vision is imperfect in that many stimuli will be indistinguishable from each other even though their wavelength compositions are very different. The number of confusions between spectrally distinct stimuli will be reduced by the presence of a third type of receptor. Three such types have been detected in many animals, ranging from goldfish (*Carassius auratus*) to primates, including humans. The vision of such animals is known as **trichromatic** vision. Although more effective than dichromatic vision, many physically distinguishable stimuli are still confused. White light, for example, stimulates all three receptors more or less equally, and it is easy to arrange a mixture of three monochromatic lights that will also stimulate the three receptors equally, and so appear white. By appropriate manipulation of their intensities, a mixture of three monochromatic lights can be made to stimulate three receptor types in almost any ratio we choose. Since the colours of objects depend upon the relative degree to which they stimulate the three receptor types, this means that such mixtures can be made indistinguishable from almost any other stimulus, irrespective of its spectral composition. The presence of four or more types of receptor, each with its own characteristic spectral sensitivity, would further improve colour vision. Such further development has certainly occurred in many birds and some reptiles.

comfort behaviour Activities that have to do with body care, including bathing, dusting, *grooming, *preening, scratching, shaking, stretching, sunning, and yawning.

communication The transmission of information from one individual to another, which is designed to influence the behaviour of the recipient. This means that *natural selection must have acted on the sender to fashion the signal, and on the recipient to detect the signal. Thus fortuitous information transfer does not count as communication. For example, when a cow (*Bos primigenius taurus*) lifts its tail prior to defecation, and the animal behind moves out of the way, we would not say that the cow is communicating with the animal, because tail-lifting has not been designed as a signal in this case.

In a simple communication system, a **source** encodes and transmits a signal, which is detected by a **receiver**, and decoded into meaningful terms. Encoding involves transformation of a message from one representation to another by operation of code rules. Thus an angry dog (*Canis*) may bare its teeth. This **signal** is the physical embodiment of a **message**, or set of **signs**, by which an animal influences the state of another animal.

In the natural environment, it is important to distinguish between **transmitted** information and **broadcast** information. Transmitted information is measured in terms of an increase in the predictability of the receiver's behaviour following activity by the source, or sender. Broadcast information is measured in terms of an increase in the predictability of the sender's identity or behaviour after a signal. Thus, broadcast information is a measure of the information obtained from a signal by an observer.

Broadcast information depends upon the state of the sender. The mapping from the sender's internal state to its communicative behaviour is an aspect of **encoding**. In other words, it is envisaged that the signal is encoded as a result of a translation of the sender's internal state into behaviour and other aspects of communication, such as *pheromone release. By the same token, the receiver translates the signal into a change in its own internal state, a process called **decoding**. The exact nature of translation depends upon the signal, the concurrent influence of other external stimuli, and the animal's current internal state. Errors in reception may occur as a result of poor detection of signals, failing to classify them correctly, and confusion between signals and irrelevant stimuli.

The effectiveness of animal signals is influenced by the physical environment, the nature of the receiver, and the influence of other signallers. Different sensory modalities are best suited to different habitats. Much depends, therefore, upon the physical nature of the habitat.

In the case of bird vocalization, for example, the two fundamental problems of transmission are attenuation and distortion. Sound distortion may result from scattering by foliage and reverberation from rocks and tree trunks. Pure tones are less subject to distortion but suffer greater attenuation through interference. However, when coding rather than transmission is of prime importance, then frequency coding is less affected by weather factors than is amplitude coding. Thus the vocalization of small birds, which cannot produce low-frequency sound, can minimize interference by being produced from a high position, well clear of obstacles.

In addition to the physical structure of the environment, the effectiveness of communication is also influenced by the nature of the receiver. A signal is effective only in so far as it is tailored to the sensory apparatus and receiver psychology of the recipient. The receiver psychology embraces the detectability, discriminability, and memorability of the signal. In addition, the effectiveness of communication will be influenced by other signallers, particularly where there are closely related species active in the same season and habitat. For example, amongst the weakly electric fish (*Gymnotus* spp.), communication takes place in murky waters by means of electrical discharges that can be pulsed up to 300 pulses/s. These fish can vary the pulse rate as a means of communication, or as part of a jamming avoidance response designed to reduce interference from other members of the species. In other words, when one fish is subject to electrical interference from another fish, it can change its pulse rate to reduce the interference.

Many animal activities undergo *ritualization during the course of evolution so that they come to serve a communication function. Any activity that is a potential source of information to other animals may become ritualized. *Displays and other forms of communication have evolved in such a way that the mechanisms that send and receive a signal are kept in tune with each other. In this way an animal is able to convey information about its internal state to another member of the same species. However, in considering communication between members of different species, it is hard to understand, in evolutionary terms, how an individual animal can benefit by informing another individual about its true state of *motivation, or about what it is likely to do next. Such an animal would gain an advantage by misleading signals leading to *deceit, and even *manipulation, of individuals of other species. This is particularly true of *anti-predator behaviour, *mimicry, and *parasitism.

Evolution of communication

The design of a communications system, whether by natural selection or by humans, should be viewed in terms of the payoff, which in the case of natural selection is *fitness. The design opportunities that present themselves to an evolving communication system are fourfold: (1) **Mutuality** occurs when both signaller and receiver benefit from the interaction. An example is the dance of the cleaner wrasse (*Labroides dimidiatus*), a *display that induces a larger host fish to open its mouth, allowing the wrasse to enter safely and remove debris and parasites. Both parties benefit from this arrangement. (2) ***Deceit** occurs when the signaller's fitness increases at the expense of the receivers. (3) ***Manipulation** by receivers can occur when the receiver obtains information about the signaller, against the interests of the signaller. For example, the courtship display of a male may attract rivals. (4) **Spite** would occur if the signaller reduced its own fitness in the process of harming another.

There are two important considerations for the success of signalling strategies. First, misleading signals must occur rarely in relation to normal signals, otherwise receivers will adjust their rules for decoding signals. The second consideration is that natural selection should act on senders to increase the efficiency of transmitting information whenever the sender's fitness is increased by the response, and to minimize efficiency whenever fitness is reduced. Thus deceit by signallers is likely to be followed by **retaliation** by receivers. Deceit occurs when signallers can take advantage of receiver's rules. Retaliation can arise by **devaluation** of the signal by the receiver, or by increased *discrimination by receivers.

community An association of plants and animals living together in a particular *habitat. Species making up a community are sometimes classified into **producers**, **consumers**, and **decomposers**. The producers are green plants that trap solar energy and turn it into chemical energy. The consumers are the animals that eat the plants, or one another, thus depending upon the plants for energy. The decomposers are generally bacteria and fungi that break down plant and animal material into a form that can be re-used by plants. Most communities have a characteristic *food chain, in which each species has a particular *niche, which is necessary for the survival of the other species.

compass An aspect of orientation, denoting the ability to head in a particular compass direction, without reference to landmarks. Animals are known to possess compasses of various types. These are based upon features of the geophysical environment, including the magnetic field of the Earth, the pattern of stars, the position of the sun, and the pattern of polarization of sunlight. These compasses are often used in conjunction with the animal's *biological clock, thus enabling compensation for the passage of time. Such time-compensated compasses play an important role in *navigation and *homing.

competition A process that occurs when two or more individuals are using the same resources, and when those resources are in short supply. Food and space are the common essential resources for which animals compete. Competition between members of the same species is termed **intraspecific** competition, and often takes the form of direct interference and *aggression between individuals. **Interspecific** competition occurs between individuals belonging to different species. It most commonly takes the form of exploitation of a resource by one species, thus denying the use of the resource to members of the other species by reducing its availability.

Competition occurs when there is overlap in the *niches occupied by different individuals. It often results in dominance of one species over another, in the sense that the dominant species has priority in the use of resources, food, space, and shelter. Subordinate species may be excluded from the use of those resources that the dominant species also uses. Consequently, it is generally true that two species with identical ecologies cannot live together in the same place at the same time. The corollary of this *competitive exclusion principle is that, if two species coexist, there must be ecological differences between them.

competitive exclusion principle The principle that two species with identical ecologies cannot live together in the same place at the same time, because of *competition between them. The corollary is that, if two species coexist, there must be ecological differences between them.

conditioning A process involved in *learning, in which an animal forms an *association between a previously significant stimulus and a previously neutral stimulus or response. In *classical conditioning an association may be formed between a significant stimulus, such as the sight of food, and a neutral stimulus, such as a flashing light. Initially, the animal responds only to the food (e.g. by salivating). If the food is presented together with the light on a number of occasions, then the animal comes to associate the light with the food, and will eventually salivate even if the light is presented alone. The response of the animal is said to have become conditioned to the light.

In *instrumental, or *operant, conditioning, the association is made between a particular response and a particular *reinforcement situation. Thus in a laboratory study, a rat (*Rattus norvegicus*) may associate pressing a lever with delivery of food. In nature, a bird may associate turning over a leaf with the discovery of an insect. In both cases the response is instrumental in obtaining food, which is said to reinforce the response. Such reinforcement may be negative. Thus a cow (*Bos primigenius taurus*) that touches an electric fence and receives a shock becomes conditioned not to touch the fence.

Conditioning is the most prevalent learning process found amongst animals, including humans.

conflict A state of *motivation in which tendencies to perform more than one activity are expressed simultaneously. At any particular moment, an animal has many different incipient tendencies, but by a process of *decision-making, one of these becomes dominant. Generally, only one tendency becomes dominant, but in certain circumstances more than one competes for dominance, and conflict arises.

Conflict has traditionally been divided into three main types:

(1) **Approach–approach** conflict occurs when two tendencies in conflict are directed towards different *goals. In such a case the animal may reach a point where the two tendencies are in

balance. However, the tendency to approach a goal generally increases with proximity to the goal. This makes approach-approach conflict unstable, because any slight departure from the point of balance, towards one goal, will result in an increased tendency towards the other, thus resolving the conflict.

(2) **Avoidance-avoidance** conflict occurs when the two tendencies in conflict are directed away from different points. Since the tendency to avoid objects generally increases with proximity to the object, movement toward either object is likely to result in a return to the point of balance. Such situations are not normally stable, because the animal can escape in a direction at right angles to the line between the two objects.

(3) **Approach-avoidance** conflict occurs when one activity is directed towards a goal, and another away from it. For example, an animal may have a tendency to approach food, but be frightened of the strange food dish. The nearer it approaches the food, the stronger the approach tendency, but the nearer it gets to the dish the stronger the avoidance tendency. The animal can reach the food, only when the approach tendency is larger than the avoidance tendency. Often there is an equilibrium point, some distance from the goal. When the animal approaches beyond this point, avoidance is greater than approach. When the animal retreats, approach is larger than avoidance. Such situations tend to be stable, because the animal is always pulled towards the equilibrium point.

Approach-avoidance conflict is by far the most important and most common form of conflict in animal behaviour. Typically such conflict is characterized by compromise and ambivalence, especially near the point of equilibrium. Irrelevant behaviour, such as *displacement activity, is also common, as are various forms of *display and *ritualization. Ritualized conflict behaviour often occurs in territorial disputes, and often forms the basis of *threat display.

conformers Animals that allow their internal state to be influenced by external factors, such as temperature, salinity, etc. To be distinguished from *regulators, which maintain their internal environment in a state that is largely independent of external conditions.

consciousness A mental state (of humans) that is sometimes attributed to animals. The main problems in doing this are: (1) consciousness, even in humans, is difficult to define; and (2) there is no accepted methodology for obtaining evidence of consciousness in animals. Consequently, we have no conception as to what consciousness in animals might involve.

conservation An aspect of wildlife management that is concerned with the preservation of endangered species. Many species are threatened with extinction as a result of human activity—some are excessively hunted and others have their *habitat destroyed as a result of agricultural or industrial development.

constraints Factors that prevent an animal from exercising behavioural options. Constraints are usually invoked in the study of animal *decision-making and animal *learning. They may include *budget constraints, or constraints on the animal's abilities.

consummatory behaviour Behaviour that brings to an end a period of *appetitive or *searching behaviour. In some cases, simply performing a consummatory act is sufficient to terminate a particular activity. For example, ejaculation brings male *sexual behaviour to an end, in many species. In other cases, achieving some *goal constitutes the consummatory situation. For example, *foraging behaviour may be terminated by finding food.

control The ability to determine the value of dependent variables. Complete controllability means that it is possible to move any state (of a system) to any other state during any finite interval of time by a suitably chosen input (i.e. by manipulation of suitable independent variables). In simple terms, when the value of a dependent variable y is determined uniquely by an independent variable x, the behaviour of y may be said to be controlled by x. When the value of y in no way influences the value of x, the control can be said to be ballistic (or open-loop). Such situations are characteristic of *fixed action patterns.

In an open-loop control system, the value of the output is likely to be affected by disturbances (changes in the mechanism due to outside influences). However, in a closed-loop system, such disturbances can be compensated for by *feedback from the output, such that the input is appropriately adjusted.

When the value of *x* is influenced by the value of *y*, a feedback situation exists and control is of the closed-loop type. Feedback exists between two variables whenever each affects the other. Positive feedback exists when the value of *y* tends to increase the value of *x*, which in turn increases the value of *y*. Positive feedback systems are inherently unstable, but they can have certain advantages in complex systems containing built-in constraints.

Negative feedback systems are characterized by the situation where the value of *x* is diminished as the value of *y* increases. The consequence is that the value of *y* therefore decreases. The value of *y* is then said to be under negative feedback control. Such negative feedback systems are common in the control of animal behaviour. Examples can be found in *motivation, *orientation, and *thermoregulation.

In ordinary terms, *a* controls *b* if, and only if, the relation between *a* and *b* is such that *a* can drive *b* into whichever of *b*'s normal range of states *a* wants *b* to be in. Thus to be a controller, *a* must have desires (or their equivalent) about the state of *b*. This means that wherever one animal controls another, that animal must have both the *motivation and the ability to exercise control. Similarly, wherever one part of an animal's internal mechanism controls another part, the former must have the ability and the instructions available to exercise control.

Control concepts such as these are important in the study of animal *communication, especially *manipulation, as well as aspects of motivation, such as *hunger and *thirst.

cooperative behaviour Behaviour in which members of a species combine in an activity, such as *hunting, *anti-predator behaviour, or care of the young, and in which there is coordination among individuals, to their mutual benefit. Cooperation among members of different species is usually called *symbiosis.

Cooperative behaviour requires good *communication among the individuals involved. Some degree of *altruism is also common, especially in situations requiring *vigilance and *alarm. The cooperative hunting of wild dogs (*Lycaon pictus*), lions (*Panthera Leo*), and hyenas (*Hyaena*) usually occurs among relatives, and can be accounted for in terms of *kin selection. Cooperation in *parental behaviour is necessary for reproductive success in many species, especially if the young need to be fed and protected by the parents.

coordination The integration of muscular movements to produce effective behaviour. Coordination requires that information be received by the *central nervous system about the position, tension, etc. of the muscles. This information is provided by internal sense organs, including muscle spindles, tendon organs, and joint receptors. These sense organs provide information about the relative positions of limbs, or other organs of movement. They also enable animals to make *reflex responses to changes introduced either by outside agents or by movement of the animal itself.

Coordination is achieved through two main processes: central control and peripheral control. The former involves a precise series of instructions produced by the *brain and obeyed by the muscles. The latter is achieved through *sense organs in the muscles that send information to the brain, and thereby influence the instructions sent by the brain to the muscles. In most cases coordination is achieved through a mixture of these two processes.

copulation That part of *sexual behaviour that is most closely associated with the fertilization of the egg by the sperm. In mammals and birds fertilization takes place inside the body of the female, while in most other animals fertilization is external. In frogs and toads, for example, the male clasps the female from above and maintains this position throughout the deposition of the eggs, fertilizing them as they emerge from the female's cloaca.

In many cases copulation is preceded by *courtship which is fairly flexible, but becomes increasingly stereotyped as progress in made towards coitus. Copulation itself is largely *reflex. Mammalian copulation is characterized by intromission, vaginal penetration by the penis. In some species ejaculation is achieved during a single intromission, while in others multiple intromissions are necessary. Ejaculation refers to the pattern of reflexes that occurs at the time of sperm emission. It is normally followed by a refractory period, during which the male refrains from copulation. This may be a matter of seconds in some species, or of days in others.

copying *See* Imitation.

cost A notional measure of the change in *fitness that is associated with an animal's behaviour, and with its internal state.

cost function A combination of the various costs, which purports to evaluate every aspect of an animal's state and behaviour in terms of its associated cost. The components of the cost function include factors associated with the behaviour occurring at a particular time, and the risks associated with the animal's internal state. For example, a bird incubating eggs incurs physiological costs in keeping the eggs warm, and costs that result from increasing hunger while it is on the nest. These costs contribute to the animal's overall fitness.

cost of changing The decrement in fitness that arises when an animal is changing from one activity to another, and receiving benefits from neither. The cost may involve loss of valuable time, expenditure of energy, or risk from predators.

courtship Behaviour that precedes the *sexual act that leads to the conception of young. Courtship serves a number of important functions, including *advertisement, mate attraction, *mate selection, mate assessment, and sexual coordination.

Attraction of possible mates is particularly important in species that are widely dispersed. For example, the females of the silk moth (*Bombyx mori*) produces windborne *pheromones that attract males. So sensitive are the antennae of the male that he is able to detect the pheromone several miles from its source. The sensory modality used in mate attraction varies with the species and its environmental circumstances. Thus birds that live in woodland tend to use *vocalization, whereas those inhabiting open spaces often use visual *display. For example, the male lapwing (*Vanellus vanellus*) performs a distinctive soaring and tumbling flight pattern which, together with his black and white plumage, makes him conspicuous from a considerable distance. In some species mates find one another as a result of *aggregation. For example, frogs and toads migrate to ponds and lakes to mate and lay their eggs.

It is important that an animal mates with a member of its own species, because hybrid mating rarely results in viable offspring. Ensuring such *reproductive isolation is an important function of *mate selection. In many species the *isolating mechanism is the behaviour that first brings mating partners together. Only members of the same species are attracted to the mating advertisement. For this reason the behaviour patterns by which animals attract mates are highly distinctive for each species. This is particularly important for closely related species for whom the potential risk of hybrid mating is greatest.

Once male and female have come together, it is often the case that one is more ready to mate than the other. Often the male is the more eager, and a function of his courtship routine is to arouse the female. Male courtship *display varies considerably from species to species. In some, such as the peacock (*Pavo cristatus*), it serves to show off his plumage to best advantage. In others, such as the wolf spider (*Pisura mirabilis*), it involves presenting the female with food. As well as arousing the female, courtship displays may serve other functions. The splendid peacock's tail may indicate the he is a healthy individual, while the spider's *courtship feeding may preoccupy the large female while the very small male copulates with her.

In cases where there is a more than one possible mate, it is important for the male to select a female that is fertile, and for the female to select a male that will make a good parent. For example, the male bullhead (*Cottus gobio*) waits in his burrow until a female swims past. Then he rushes out and seizes her by the head in his jaws. Only females that are fully sexually mature remain quiescent and allow themselves to be pulled by the head into the male's burrow where spawning will take place. Immature females respond to the male's bite as if it were an attack, and wriggle free. Thus the male tests the sexual state of the female by apparently aggressive behaviour, ensuring that he mates only with a fully mature one. Females often test males during courtship, by being initially coy, and forcing the male to put a lot of effort into his overtures. In the smooth newt (*Triturus vulgaris*), for example, courtship takes place under water, where the male actively displays to a relatively quiescent female. The male, therefore, uses up oxygen much faster than the female, and eventually may be obliged to break off the courtship and ascend to the surface to breath. It has been suggested that females test the physiological fitness of the male by prolonging the courtship. If the male has to breathe before fertilizing the eggs, he may have lost his chance, especially if there are other males around.

If mating is to lead to conception, it is important that both male and female play their part correctly. The moment when the male attempts *copulation must coincide with the time at which the female reaches her fully responsive state. Timing is particularly important in those species where fertilization is external, because any delay may allow the gametes to disperse before they can meet.

courtship feeding Presentation of food by one partner to the other during *courtship. Among birds, such as the black-headed gull (*Larus ridibundus*), the female often begs for food in the manner of a juvenile. When courtship feeding extends into the incubation period, the food may be a substantial contribution to the female's requirements.

Courtship feeding also occurs among invertebrates. The male wolf spider (*Pisura mirabilis*), for example, is smaller than the female and in some danger of being eaten by her. He captures a suitable prey, wraps it in silk, and presents it to the female during courtship. If the female accepts the food, the male is able to approach her and copulate unmolested.

crypsis A form of *camouflage, in which the animal resembles part of the environment. Some scientists exclude cases where the predator can see the cryptic animal, but fails to recognize it as being edible. Such cases of camouflage in which the animal resembles another animal, or part of a plant or stone, are regarded as *mimicry.

cue strength The combination of external stimuli that give rise to a particular activity. For example, the colour markings of the cichlid fish (*Haplochromis burtoni*) are additive in their effect in eliciting aggression in a rival fish. They combine to indicate the potential threat. The magnitude of this combination is the cue strength for threat.

cultural behaviour Behaviour that has been passed from one generation to another by non-genetic means. *Evolution occurs primarily as a result of *natural selection, and genetic inheritance of acquired characteristics is not possible. However, information can be passed from parent to offspring through the processes of *imprinting and *imitation.

*Sensitive periods of learning occur in the early life of many animals, and during such periods they often learn from their parents. Some songbirds remember the parental *song, provided they hear it during the sensitive period (about ten to fifty days old), and are able to sing it in later life. The tendency to copy the parental song leads to regional variation. Populations separated by only a few miles may have different dialects. Animal dialects represent an elementary form of *tradition. Other forms of traditional behaviour include *migration routes and feeding habits.

*Imitation provides another mechanism for cultural exchange in animals. Animals may copy each other as a result of simple *social facilitation, *learning by observation, or true imitation. Whether the last of these indicates *intelligent behaviour is a matter of controversy.

curiosity A *voluntary form of *exploratory behaviour, as opposed to *reflex exploration, *startle, or *play. Curiosity and *fear seem to be related, in that animals will sometimes approach objects or other animals of which they are usually frightened. For example, Thomson's gazelle (*Gazella thomsoni*) and wildebeest (*Connochaetes* sp.) often approach and stare at predators such as the spotted hyena (*Crocuta crocuta*), cheetah (*Acinonyx jubatus*), and even lion (*Panthera leo*). It has been suggested that curiosity is motivated by a desire to learn.

daily routine A day-long pattern of behaviour that tends to be repeated day after day. Environmental changes between night and day affect animals both directly and indirectly. Thus there may be changes in food availability and in numbers of predators, which are brought about by changes in light intensity, temperature, etc. In adjusting to the differences between night and day, the animal adopts a daily routine that is made up of many different aspects of behaviour fitted together to form a pattern that tends to be repeated day after day.

When individual activities are studied, they tend to follow a *circadian rhythm, which is partly due to the fact that the activity must fit in with other aspects of the daily routine, and partly due to the influence of the *biological clock.

Daily routines enable animals to make the best use of their time and to exploit opportunities. Thus the European kestrel (*Falco tinnunculus*) is a diurnal predator that specializes in preying upon small mammals. They catch prey throughout the day, but do not immediately eat what they catch. They tend to cache surplus prey throughout the day, and retrieve them at dusk. This routine enables the kestrel to make the most of available prey throughout the daylight hours without spending too much time eating the prey.

Darwin, Charles Robert (1809–82) The founder of the theory of *evolution by *natural selection, which received its most famous expression in his book *On the Origin of Species by Natural Selection* (1859). Darwin wrote a number of other books that have implications for the study of animal behaviour. The most important were *The Descent of Man and Selection in Relation to Sex* (1871) and *The Expression of the Emotions in Man and Animals* (1872).

Darwinian fitness A measure of capacity to produce offspring. In any population of animals there is variation among individuals and, consequently, some have greater reproductive success than others. Individuals that produce a higher number of viable offspring are said to have greater Darwinian fitness.

The *fitness of an individual depends upon its ability to survive to reproductive age, its success in mating, the fecundity of the mated pair, and the probability of survival to reproductive age of the resulting offspring. The number of offspring must be counted at the same stage of the life cycle. The fitness of an individual must take its age into account, because the potential for reproduction changes with age.

death feigning A form of *defence in which the animal becomes immobile, as if dead. Under natural conditions this usually occurs at the closing stages of a *predatory encounter. In response to being grasped by the jaws of a predator the prey becomes immobile, and this behaviour can be induced in the laboratory by restraining the animal with the hand, as in animal *hypnosis. The *survival value of death feigning lies in the fact that predators do not attack dead prey, and do not always immediately eat animals that they kill. Foxes (*Vulpes* sp.) may kill a number of prey, some of which may be buried. It has been reported that ducks (Anatidae) buried in this way have subsequently escaped.

deceit A form of *evolutionary strategy, involving *communication. Deceit occurs when the signaller's *fitness increases at the expense of the receiver's. For example, many ground-nesting birds feign injury when their nest is approached by a predator, such as a fox (*Vulpes* sp.). The bird moves away from the nest trailing an apparently broken wing, thus luring the fox away from the nest. This type of *distraction display is not usually regarded as a deliberate ploy on the part of the individual animal, but rather an instinctive reaction to an approaching predator. However, some instances of deceitful behaviour, particularly amongst primates, are regarded by some scientists as indicating *intention, or *cognition, on the part of the individual animal.

A common form of deceit is *mimicry in which one animal (the mimic) resembles another animal (the model), so that the two are confused by a predator. Often the model is avoided by the predator, and this avoidance is extended to the mimic. For example, birds that avoid wasps (*Vespula* spp.) that have warning *coloration, also avoid the similar but harmless hoverflies (*Syrphus* spp.). For a deceitful strategy to be successful, it is important that the misleading signals occur rarely in relation to normal signals. Otherwise receivers will adjust their rules for decoding signals. Thus when a fox (*Vulpes vulpes*) encounters an apparently crippled ringed plover (*Charadrius hiaticula*), and is deceived, it adopts the rule 'stalk the bird'. However, if the fox were to be unsuccessful too often, it might adopt the rule 'look for the nest'.

decision-making The process of changing from one activity to another, where there is incompatibility between the two activities. The term 'decision-making', in animal behaviour studies, carries no implications about the mechanisms employed, whether reflex or deliberative. However, it is generally assumed that decisions are made between activities relating to different *goals, as opposed to choice between alternative routes to a single goal.

A distinction between design and execution is important in the study of decision-making in animals. Generally, the design has been carried out by *natural selection during the process of evolution, and the execution is the business of the individual animal. The design specifies what the animal 'ought' to do, but it is up to the individual to do the 'right thing'. It is generally assumed that the latter is achieved by 'rules of thumb', except perhaps in species whose *intelligence enables some evaluation of costs and benefits to be carried out by the individual animal.

In considering design, what an animal ought to do in a given situation, we have to remember that the various possible activities differ in their consequences and have different costs and benefits attached to them. It is assumed that the animal is designed by natural selection to behave in such a manner that the greatest net benefit is attained. This approach makes possible the application of conventional decision theory, which is based upon assumptions about *rationality and *utility.

defence Any behaviour that reduces the chances of one animal being harmed by another animal. Defensive adaptations may be static, such as the spines of a hedgehog (*Erinaceus* sp.), or active, such as running away. But almost all defence has a behavioural component. Thus, when attacked, the hedgehog rolls into a ball.

While the most common systems of defence are adaptations against predators, animals may also have defences against parasites and against other members of their own species. For example, as well as having a social function, mutual *grooming helps defend against ectoparasites. Similarly, *territorial behaviour helps defend against rivals.

Defence against predators may be primary or secondary. Primary defences operate regardless of whether or not there is a predator in the vicinity. They reduce the probability that a predator will encounter an animal. They include *camouflage and *mimicry, group living and some forms of *social interactions, and *symbiosis.

Secondary defences operate only after an animal has detected a predator. They increase the chances that the animal will *escape from the encounter. They include withdrawal, flight (escape), bluff (*deimatic display), *death feigning, deflection of attack, and retaliation.

Defensive behaviour incurs costs. Escape from one type of predator may make an animal more vulnerable to another. Camouflage often necessitates periods of motionless during which the animal cannot feed. Sticklebacks (*Gasterosteus aculeatus*) are best protected from predators if they are muted in colour, but the males are more attractive to females if they have conspicuous red bellies. Hence, the colour of male sticklebacks in any population will depend upon the relative strengths of predation pressure and female preference. Thus defensive behaviour is a compromise between the requirements of defence and those of other essential systems.

deimatic display An evolutionary strategy in which an animal adopts a *display designed to scare off a predator. For example, the caterpillar of the hawkmoth (*Leucorampha* sp.) normally rests upside down beneath a branch or leaf. When disturbed, it raises and inflates its head, the ventral surface of which has conspicuous eye-like marks, and the general patterning of which resembles the head of a snake. Many moths and butterflies have eye-spots on their wings, which they reveal suddenly when disturbed, with the possible effect of frightening the predator.

demand function A concept borrowed from economics to express the relationship between the price and the consumption of a commodity. For example, when the price of coffee is increased and people continue to buy the same amount as before, demand is said to be inelastic. When the price of fish is increased, and people buy less than before, demand is said to be elastic.

Exactly analogous phenomena occur in animal behaviour. If an animal expends a certain amount of energy on a particular activity, then it usually does less of that activity if the energy requirement is increased. The elasticity of demand functions in animals gives an indication of the relative importance of the various activities in the animal's repertoire, and demand functions are closely related to *behavioural resilience.

development Synonymous with *ontogeny which is concerned with the history of the individual from conception to death, and with the roles of *genetics, *maturation and *learning in shaping the animal's life history.

dialect *Vocalization of a population of animals that differs from that of another population of the same species. The occurrence of dialect is widespread in bird *song, and has also been discovered in vocalizations of other types of animal, such as the mating calls of the Pacific tree frog (*Hyla regilla*) and of various marine mammals.

The formation of dialects requires some ability to acquire elements of the vocal repertoire through *learning and *imitation. It is generally seen as an aspect of *imprinting, whereby the juvenile animal learns some of the characteristics of its own species. Dialects give a certain distinctiveness to local populations, and these may diverge from each other if interbreeding is impaired by ecological or geographical isolation (*see* ISOLATING MECHANISMS).

dilution effect By joining a group, an individual dilutes the impact of a successful attack by a predator, because there is a chance (depending on group size) that another animal may be the victim.

discrimination Making different responses to different stimuli. In **successive discrimination** the stimuli are presented to the animal one after another. For example, the freshwater polyp, (*Hydra* sp.), stings prey that emit the biochemical glucothione, but does not sting objects that do not emit this chemical. *Hydra* spp. discriminates between the presence and absence of the chemical when objects are touched successively.

All animals are capable of successive discrimination, as is shown by the fact that they behave differently to different stimuli. Often, the discrimination is rather crude, as is shown by the phenomenon of *sign stimuli. For example the European robin (*Erithacus rubecula*) discriminates between birds that enter its territory that have a red breast and those that do not. Those with a red breast are attacked. Indeed, even a bunch of red feathers (no bird body present) will be attacked, this indicating the crude nature of the stimulus. However, many animals can learn to refine their discrimination powers, a process that is called **discrimination learning**. Pigeons (*Columbia livia*), for example, have been trained to discriminate between shades of green (wavelength around 550 nm) differing by only 10 nm.

In **simultaneous discrimination** studies, different stimuli are presented to the animal at the same time, so that the animal has a *choice. Such studies have shown that simultaneous discrimination is partly a matter of *attention. Thus, if pigeons are trained to discriminate among stimuli of different colours, or among stimuli of different shapes, and they are then required to learn a new discrimination in which the stimuli differ in both colour and shape, they show transfer of learning. Those that previously learned a colour discrimination learn the new discrimination more quickly than those that previously learned a shape discrimination, if colour is the relevant (i.e. rewarded) cue. Whereas those that previously learned a shape discrimination do better if shape is the relevant cue in the second task. In other words, in the first task the animals learn to discriminate among particular stimuli, but they also learn that the must pay attention to the colour (or shape) of the stimuli. This prior learning may benefit them in the second task.

disinhibition The removal of inhibition. There are two types of disinhibition: *motivational disinhibition and *Pavlovian disinhibition.

dispersion The distribution in space of individuals within their natural *habitat. The dispersion pattern of a species is determined partly by topological factors, and partly by the behaviour of individuals towards each other. Topological factors may include the distribution of nest sites, sleep sites, or food sources, features related to variation in microclimate, and physical barriers to dispersion.

Apart from irregularities due to topological factors, the patterns of dispersion observed in nature are either random, clumped, or regular. Random dispersion is rare, although it can be found in various marine molluscs (*Mollusca*), where it may be a defensive measure against the *searching strategies of predators.

Clumped distributions are common, especially in species that tend to form family groups, herds, etc. Clumped patterns often indicate the defence of some resource, such as *territory, females, etc. For example, during the rut the red deer (*Cervus elaphus*) of Scotland range over large areas of unwooded hills. Each stag attempts to gather together a number of hinds, which he defends against other stags. This behaviour results in a constantly changing clumped pattern of dispersion.

Regular dispersion patterns often result from situations in which there are both clumping and spacing tendencies, as in *schooling in fish, and some territorial situations. Ideally, a single territory will be circular, since this shape has the largest area for the boundary to be defended. When the density of territories is very high, we might expect them to become hexagonal, because hexagons pack together while retaining a large area for a given boundary length. Hexagonal territories have been observed in the Mozambique cichlid (*Tilapia mossambica*). These fish excavate breeding pits at the centre of their territory. Territories not in contact with others are generally circular and slightly larger than the norm.

displacement activity Apparently irrelevant activity that occurs in situations of *conflict or thwarting. Displacement activities often occur in aggressive encounters, where the individual has simultaneous tendencies to attack and to flee. For example, fighting cockerels (*Gallus domesticus*) may briefly turn aside and peck at the ground as if feeding.

Because displacement activities often occur at the equilibrium point in an aggressive or sexual encounter, they provide potential information to the other participant. When *communication of this type is advantageous, natural selection will tend to make the behaviour more reliable and efficient as a conveyor of information. The evolutionary process by which this comes about is termed *ritualization. By means of ritualization, displacement activities can become incorporated as part of normal aggressive or courtship *display. For example, ritualized displacement preening occurs during courtship in many species of duck (Anatidae). It tends to be more stereotyped than normal preening, and to be directed towards especially conspicuous feathers.

display A stereotyped motor pattern involved in animal *communication. Displays are largely genetically determined and specific to each species. Related species often have similar displays. For example, the courtship displays of fiddler crabs (*Uca* spp.) consist of rhythmic waving of the enlarged claw, accompanied by body movements, so that the display looks like a dance. In the Philippine species *Uca zamboangana*, a series of vertical waves of the claw is accompanied by raising and lowering the body. In the South American species *Uca pugnax rapax*, the enlarged claw moves outwards and upwards in three jerks and descends smoothly. The display is always stereotyped within a species. It would not, otherwise, serve as a reliable means of communication.

Display postures often show off distinctive colour patterns, weapons, or other physical characteristics. Thus, smiling in humans is a universal display that is stereotyped, both in its detail and in the situations in which it occurs. Smiling is made more effective by the distinctive coloration of the lips.

Many displays have evolved from simple behaviour patterns through a process of *ritualization that makes them more stereotyped, conspicuous, and reliable as a means of communication. Some have evolved from *intention movements and *displacement activities. Others have evolved from *defence behaviour and from physiological responses involved in mild stress.

distraction display An injury-feigning display that functions to lure a predator away from a nest, or other location of importance to the animal. For example, if the sandpiper (*Ereunetes mauri*) is disturbed by a ground predator while incubating, it may leave the nest and act as though injured, often trailing an apparently broken wing. When it has lured the predator a safe distance from the nest, it regains its normal behaviour and flies away. Thus distraction displays are a form of *deceit.

distress A state of motivation relating to *stress that the animal is unable to cope with. An animal may be in distress as a result of physical environmental factors, such as excessive cold or heat, or as a result of *social interactions or *predator avoidance.

disturbance effect An effect of animals feeding in groups and disturbing prey. For example, undisturbed shrimps (*Corophium* spp.) live with their tails sticking out of the mud surface, but when disturbed vanish into the mud. Redshank (*Tringa totanus*), feeding close together, are likely to disturb each other's prey, to their mutual disadvantage.

domestication The process by which humans have structurally, physiologically, and behaviourally modified certain species of animals by maintaining them in, or near, human habitations, and by *selective breeding. Domestication is designed to suit human objectives, which may relate to economic performance, such as docility, efficient maternal care, high fertility, longevity, efficient food conversion, and increased production of materials such as wool, milk, or meat. Other objectives include ornamentation, as is the case with some fish, birds, and dogs, or entertainment, as is the case with fighting cocks, dogs, and bulls.

dominance A feature of *social organization in which some individuals acquire a high status, usually as a result of *aggression, while other individuals retain a low status. Dominance relationships were first noticed in flocks of domestic fowl (*Gallus gallus domesticus*), in which dominant individuals tend to peck subordinate individuals when they come within range. In a stable flock, individuals learn to recognize each other and a 'peck-order', or dominance hierarchy, becomes established.

Dominance relationships are widespread in the animal kingdom, and have certain features in common in many species. Dominant individuals tend to use their status to gain priority in access to resources, such as food, roosting sites, etc. Subordinate individuals show typical *appeasement behaviour, sometimes without any sign of fear. The dominant animal may simply supplant the subordinate at a feeding site, as a matter of routine. In many species, dominant males have priority in access to females, and perform most matings. Among breeding northern fur seals (*Callorhinus ursimus*) the dominant males hold territories that they defend against rivals day and night for a period of up to two months without food. As a result of fighting, the mortality rate among dominant males is three times that of females.

dormancy A period of quiescence, by means of which animals avoid unfavourable conditions. Summer dormancy, or *aestivation, enables animals to avoid extremes of climate, especially excessive heat or drought. It is most common in desert-living species, but is not confined to them. Winter dormancy, or *hibernation, enables animals to avoid being active during extremely cold conditions. True hibernation should be distinguished from other types of dormancy. True hibernators have *thermoregulation that allows the body temperature to fall to that of the surrounding air, maybe as low as 2°C. True hibernation is found in some mammals and birds. Other types of dormancy (without thermoregulation) occur in many species, especially at night, when it may be dangerous to be active. Thus many insects and fish become dormant at night, usually hiding in a safe place. Many vertebrates may appear to *sleep at night, showing a heightened threshold of arousal. However, true sleep is defined in terms of the characteristic electrical activity of the brain. This occurs only in birds and mammals.

dreaming During human sleep there are periods when the sleeper makes rapid eye movements. Physiological studies show that the electrical activity of the brain has a characteristic pattern during these phases. Paradoxically, this pattern is similar to that associated with wakefulness, although it occurs during deep *sleep from which the person is not easily aroused. People that are woken up during these **active sleep** phases usually report that they were dreaming, which they do not if woken during **quiet sleep**.

Many sleeping mammals show rapid eye movements during sleep, and these are accompanied by physiological activity similar to that found in humans. On the basis of this evidence, many scientists are willing to agree that such animals experience dreams akin to those of humans. Some are of the opinion that animals have only episodic dreams, whereas humans also have narrative dreams. However, since animals cannot report their dreams, their nature remains a matter of conjecture.

drinking Taking in water by mouth to quench *thirst. Many aquatic animals take water in through the mouth, but this may play no role

expression is an important part of *social interaction, and the more highly social species can therefore be expected to have more sophisticated emotional exchanges.

emulation A form of apparent *imitation, in which an animal duplicates the results of another animal's behaviour (i.e. achieves the same *goal), but using a different method.

endocrine system A system of (endocrine) glands that releases *hormones into the blood of vertebrates. Each hormone has specific functions and often affects specific target organs. The endocrine glands turn their secretions on and off at appropriate times. The control of hormone secretions is achieved either by direct nervous action, or by the action of other hormones.

energy The capacity for doing work, where work is a measure of the change of state of a system. The mechanical energy changes that are involved in *locomotion depend upon muscular work. The capability for this derives ultimately from the food that the animal eats. When food is digested, energy is released as the chemical constituents of the food change their state. Some of this energy is lost in the form of heat and in the excretion of the waste products of digestion. Similarly, the food chemicals that are absorbed following digestion are further changed in the process of metabolism. Some energy is again lost in heat and excretion, and some is made available for muscular movement and other bodily processes.

The amount of energy that an animal requires depends upon its way of life, and this is influenced, in turn, by the energy available in the environment. The food of herbivores has low energy content compared with that of carnivores, with consequent effects upon their lifestyle and behavioural organization. Energy *budgets, similar to those of economists, can be calculated, with energy taking the role of money. Animals earn energy by *hunting, *foraging, and *feeding, and spend it on other activities.

The term 'energy' is sometimes used in the context of *motivation, but this is misleading. Since energy is a concept of capacity, it cannot be a causal agent. Behaviour can be caused in such a way that certain energy changes are consequent, but energy, whether mental or physical, cannot drive behaviour.

escape A form of *defence behaviour, which may occur as soon as a predator is detected, or only when the predator attempts capture. For example, rabbits (*Oryctolagus cuniculus*) run as soon as a predator is detected, while hares (*Lepus europaeus*) may remain motionless, in *camouflage, until detected by a predator, and only then make their escape.

Some animals escape to a prepared retreat, such as the rabbit's burrow, which often has a second exit. Anemone fish (*Amphiprion* spp.) retreat into the tentacles of their anemone, from which they gain protection. Some animals, like the tortoises (*Testudinidae*) have a mobile retreat.

Evasive manoeuvres during escape are common. Thus hares, and ptarmigan (*Lagopus* sp.), execute sudden changes of direction during flight from predators. Flying fish (Exocoetidae) make prolonged jumps out of the water, and so disappear from the predator's view. Many insects, and some frogs (Anura), exhibit flash *coloration. Thus grasshoppers (*Trilophidia* spp.) have conspicuous coloration while escaping, then they suddenly assume a stationary cryptic posture and apparently disappear.

Many invertebrates have a stereotyped escape *reflex, triggered by giant nerve fibres that carry a very fast message to all the muscles involved, so that they contract suddenly and simultaneously. Examples are the withdrawal reflexes of snails (Gastropoda) and the tail flip of prawns (*Decapoda*) that propels the animal backwards.

ethogram A pictorial representation of the frequency with which one activity follows another (usually depicted in terms of the thickness of the relevant arrow). It serves as an inventory for the behaviour of a particular species.

ethology The study of animal behaviour, which attempts to combine causal and functional types of explanation. The former is concerned with the way in which proximate causal mechanisms combine to control the behaviour of animals. These may concern *orientation, *learning, and *motivation.

The alternative form of explanation offers hypotheses that aim to show how *natural selection has, in the past, acted as a designing agent in shaping the *evolution of behaviour. Such explanations account for behaviour in terms of its *function. Traditionally, ethologists have sought to combine observation of behaviour in the wild with experiments designed to illustrate the function of the behaviour.

For example, Karl von *Frisch maintained, in contrast to the prevailing view at the time (1914), that honey-bees (*Apis mellifera*) showed *coadaptation with respect to the structure of flowers, and were therefore likely to possess *colour vision. He showed by experiment that this was indeed the case.

evolution Origination of species' characteristics by development from earlier forms. Now acknowledged to be due to the process of *natural selection. Evidence for evolution comes from the fossil record, comparison of present-day species, the geographical distribution of species, and some observations of evolution in action. Animal behaviour can only fully be understood in terms of its evolutionary history, and in terms of the role it plays in survival and reproduction (more precisely the *inclusive fitness) of the animal. However, the evolution of animal behaviour is difficult to study directly, as there is no fossil record, although something can be learned by comparison of species. Instead, biologists have developed a body of evolutionary theory that enables them to develop hypotheses concerning the evolution of behavioural traits.

Evolutionary biologists are interested in explaining how a state of affairs observed today (such as the typical behaviour of a certain species) is likely to have come about as a result of evolution by natural selection. To account for the establishment of a particular genetic trait, they imagine a time before the trait existed. Then they postulate that a rare gene arises in an individual, or arrives with an immigrant, and that individuals carrying the gene exhibit the trait. They then ask what circumstances will favour the spread the gene through the population. If the gene is favoured by natural selection, then the individuals with genotypes incorporating the gene will have increased fitness. The gene may be said to have invaded the population. To become established a gene must not only compete with the existing members of the gene pool, but must also resist invasion by other mutant genes. It is as if the genes develop a strategy to increase their numbers at the expense of other genes. Thus an *evolutionary strategy is a passive result of natural selection that gives the appearance of a ploy employed by genes to increase their numbers at the expense of other genes. For example, a gene that makes an insect unpalatable to predators may not spread in the population, because by the time the predator discovers that the prey is distasteful, the insect is dead. However, if the gene also provided the insect with distinctive warning *coloration, then the predator would more easily learn to avoid other insects carrying the gene, and the gene would increase in frequency in the gene pool. An *evolutionarily stable strategy is a strategy that cannot be bettered by any feasible alternative strategy, provided sufficient members of the population adopt it. Such strategies are resistant to invasion.

The survival of a trait within a population depends upon the extent to which the trait contributes to *reproductive success, which depends partly upon the selective pressures inherent in the environment. A number of features of the environment could jeopardize reproductive success by leading to the death of the parent by starvation, predation, failure to breed as a result of *competition for mates or nesting sites, failure of the young to survive due to lack of *parental care, food, or protection from predators.

However, there is a form of natural selection, called *sexual selection which depends upon the advantage that certain individuals have over others of the same sex and species solely in respect of reproduction. There are two ways in which a male can gain an advantage over other males. First, they can compete directly with one another by *fighting, or by some type of ritualized combat. This is sometimes called **intrasexual selection** (selection within a sex) or **male rivalry**. Secondly, males can compete indirectly in attracting females by special displays and adornments, sometimes called **intersexual selection** (selection between sexes) or **female choice**. Thus male peacocks (*Pavo cristatus*) develop enormous ornamental tails that are attractive to females. Females prefer males with good tails, and males benefit from this preference, even though the tail is costly to produce and maintain, and is likely to hinder escape from predators. Thus different driving forces in evolution may work against each other.

evolutionarily stable strategy (ESS) An *evolutionary strategy that cannot be bettered by any feasible alternative strategy, provided sufficient members of the population adopt it. Often, the best strategy for an individual depends upon the strategies adopted by other members of the population, and the resulting ESS may be a mixture of a number of strategies. For example, two participants in a contest may not have the same choice of strategies, or prospective payoffs. In contests between male and female, old and young, or the owner of a resource and a non-owner, an asymmetry may be perceived beforehand by the contestants, and will usually influence the choice of action. Such situations are usually analysed in terms of *game theory (*see* also ASYMMETRIC GAMES).

evolutionary strategy A passive result of natural selection that gives the appearance of a ploy employed by genes to increase their numbers at the expense of other genes (*see* EVOLUTION). An evolutionary strategy is not a strategy in the cognitive sense, but a theoretical tool employed by evolutionary biologists.

excretion The elimination of waste matter from the body by defecation and urination. In most animals this is an automatic process, but in some mammals it has become ritualized, and plays a role in *communication. Marking territory by defecation occurs in various carnivores; the faeces contain important *pheromones. *scent marking by urination also occurs in many mammals.

explicit knowledge *See* Knowledge.

exploratory behaviour A form of *appetitive behaviour which may or may not be aimed at a particular commodity or environmental situation. It is shown by animals *searching for food, nest sites, etc., and when the commodity or situation is arrived at, the exploration ceases. This type of exploratory behaviour is a form of *goal achieving behaviour. Other forms of exploratory behaviour do not seem to be aimed at a specific goal. This is particularly true of responses to *novelty. Rats (*Rattus norvegicus*), for example, routinely explore novel foods, but will usually sample them in only very small quantities. This is part of their normal process of *food selection. It is not terminated by the discovery of novel food, or any other obvious goal.

Many animals seem to explore simply for its own sake. Rats explore unknown terrain, and monkeys (Simiae) often explore novel objects manually. Some animals will work for the chance to explore novel situations, and this may be an aspect of their *curiosity.

extinction The process by which learned behaviour patterns cease to be performed when they are no longer appropriate. *Learning occurs when a novel stimulus is accompanied by an event of consequence to the animal, or when the animal's own behaviour is followed by some event of consequence. When the relationship between the stimulus or behaviour is no longer followed by relevant consequences, then extinction occurs and the animal gradually ceases to make the learned response. For example, if an animal discovers food in a particular place, it may learn to visit that place on future occasions, provided it finds food there at least some of the time. If the animal no longer finds food, it will visit the place less and less frequently, until the learned behaviour is extinguished and the behaviour ceases entirely.

Extinction is not simply a matter of lack of *memory. Relearning after extinction is usually much more rapid than the original learning, which suggests that the process of extinction does not abolish the original learning, but somehow suppresses it. Further evidence for this conclusion comes from the phenomenon of **spontaneous recovery**, by which a response that has been extinguished recovers its strength with rest. If an extraneous stimulus is presented during extinction, the response increases—a phenomenon called *Pavlovian disinhibition—which implies that the process of extinction is an active inhibitory process.

F

facial expression A form of *communication in which the muscles of the face are deployed to send visual signals. Only in mammals have structures of the face, such as the lips, cheeks, eyebrows, and attached musculature, differentiated sufficiently to allow a rich repertoire of facial expression. The greatest differentiation is found in the higher primates, including humans. These species have adopted a diurnal way of life, and have evolved good *vision.

Facial expressions are compound patterns consisting of a number of **action units**. These are thought to have evolved from primitive functional units, such as those connected with *fear, *vigilance, *intention movements, and protective responses. For example, responses to danger include opening the eyes, raising the eyebrows, and dilating the pupils, all of which increase receptivity to stimuli. When a mammal pays attention to external stimuli it performs an *orienting response, in which eyes and ears are focused upon the stimulus. A mammal may protect its sense organs from noxious stimuli by closing the eyes and flattening the ears. Mammals may make intention movements, such as opening the mouth to bite, and putting out the tongue to lick.

During the evolution of facial expressions amongst the primates, there has been considerable *ritualization of the action units. This is shown by differing degrees to which the units are exaggerated in different species. Whereas humans frown more deeply than apes (Pongidae) and monkeys (Simiae), chimpanzees (*Pan troglodytes*) have broader grins. Humans raises their eyebrows, and macaques (*Macaca* spp.) raise their eyebrows and flatten their ears, but apes hardly do either of these things. All primate facial expressions, with the notable exception of lip-smacking, occur in humans. On the other hand, two human *displays—disgust or contempt, and surprise—do not have obvious ritualized counterparts in other primates. Both patterns do occur in their primitive form, but in humans they are highly ritualized.

fear A state of *motivation that is aroused by certain specific stimuli, and normally gives rise to some form of *defence or *escape. Fear-provoking stimuli may be *sign stimuli that are responded to without prior experience, or they may result from *conditioning.

Fear is usually accompanied by physiological changes, such as increased rate of heart beat and increased adrenaline levels. Many animals have characteristic fear *displays such as pilo-erection and changes in *coloration. In mammals, especially primates, there may be changes in *facial expression. *Alarm calls and *death feigning are also common responses to fear.

Extreme or prolonged fear can induce *stress and anxiety neuroses. These may result from the *dominance of one animal over another, especially in captivity. In nature, fear is sometimes accompanied by *curiosity and a tendency to investigate. For example, upon sighting a cheetah (*Acinonyx jubatus*), Thompson's gazelle (*Gazella thomsoni*) may run up from a distance of several hundred meters, and approach to within 50–80 m. All the gazelles have their ears cocked and stare at the cheetah, occasionally uttering an alarm snort.

feedback An important aspect of the *control of behaviour. When the value of a dependent variable *y* is determined uniquely by an independent variable *x*, the behaviour of *y* may be said to be controlled by *x*. When the value of *y* in no way influences the value of *x*, the control can be said to be ballistic (or open-loop). Such situations are characteristic of *fixed action patterns.

When the value of *x* is influenced by the value of *y*, a feedback situation exists, and control is of the closed-loop type. Feedback exists between two variables whenever each affects the other. Positive feedback exists when the value of *y* tends to increase the value of *x*, which in turn increases the value of *y*. Positive feedback systems are inherently unstable, but they can have certain advantages in complex systems containing built-in constraints.

Negative feedback systems are characterized by the situation where the value of *x* is diminished as the value of *y* increases. The consequence is that the value of *y* therefore decreases. The value of *y* is then said to be under negative feedback control. Such negative feedback systems are common in the control of animal behaviour. Examples can be found in *motivation, *orientation, and *thermoregulation.

feeding All the activities that are involved in obtaining, handling, and ingesting food. These range from *predatory behaviour to grazing. Different species employ different strategies, in accordance with their ecological *niche. The food of some animals, such as marine filter feeders and some vertebrate herbivores, is relatively uniformly distributed, and feeding does not require a lot of effort, although it may take up a lot of time. When food is patchily distributed, the animal must employ active *searching and *foraging strategies.

Many animals show a high degree of *food selection.When food is plentiful, the most profitable food may be selected. Thus deer (Cervidae) may walk through a field of clover to reach another field that has been fertilized, and has more luxurious vegetation. Redshank (*Tringa totanus*), foraging for the marine ragworm (*Nereis* sp.) on the sea shore, ignore the smaller worms and select only the large ones. In general, the feeding mechanisms of each species are designed to enable it to feed in the most profitable manner, without undue waste of energy in handling the food, or in ingesting food that is not worthwhile. At the same time, the feeding animal has to contend with a number of potential hazards and dangers. Poisonous animals and plants must be avoided. Domestic cattle reject food that has a bitter taste, and birds quickly learn to avoid conspicuous insects on the basis of their taste. Many animals exercise caution in sampling novel food, and may take a little, and then wait a few hours before taking more. The sign of other animals feeding often has a facilitatory effect, and it is probable that some degree of insurance against poisoning is obtained in this way. Feeding animals must also be alert to the possibility of being surprised by predators, and it is notable that many birds and ruminants look up repeatedly while feeding. Such animals show more *vigilance when a predator has been seen in the vicinity.

The *motivation for feeding behaviour is based in part upon the principle of *homeostasis. As the animal uses up energy and nutrients, various imbalances occur in the animal's internal environment, and these are registered by the brain. Deprivation of food results in a state of *hunger, and establishes feeding as a high-priority activity. However, *appetite does not depend on hunger alone, but may be aroused by the sight of food, or the opportunity to obtain a particular food. Animals learn when and where food is likely to be found, and distribute their feeding effort accordingly. Appetite may be depressed by high ambient temperature, unpalatable food, or other unfavourable circumstances. Many animals feed on a routine basis, established by their internal *biological clock, and time of day can be an important stimulus to appetite. In addition, decisions about feeding are influenced by the animal's assessment of the competition from other individuals, the dangers of predation, etc. *opportunity is an important determinant of feeding behaviour, especially in *hoarding animals.

fighting An aspect of *aggression, in which combat occurs during disputes over some resource, such as food, territory, or mates. Fighting does not normally occur between members of different species, except in cases of mistaken identity.

Fighting usually takes the form of *ritualized combat. For example, the oryx (*Oryx gazella*) has sharp, pointed horns that could inflict mortal wounds. These may be used in *defence against predators, but in contests among oryx, the horns are used in a purely ritualized manner. The rivals lock horns and wrestle. It is against the rules to stab a rival in the side. Similarly, rattlesnakes (*Crotalus* spp.) settle their differences with ritualized trials of strength in which one attempts to pin the other to the ground. They do not use their poisonous bite against rivals. (*See* Figures 3 and 4.)

Fighting is usually preceded by *assessment, in which the antagonists assess each other's fighting potential, by looking at body size, weapons, etc., and by detecting *pheromones that may give clues to the degree of *fear and commitment of the antagonist.

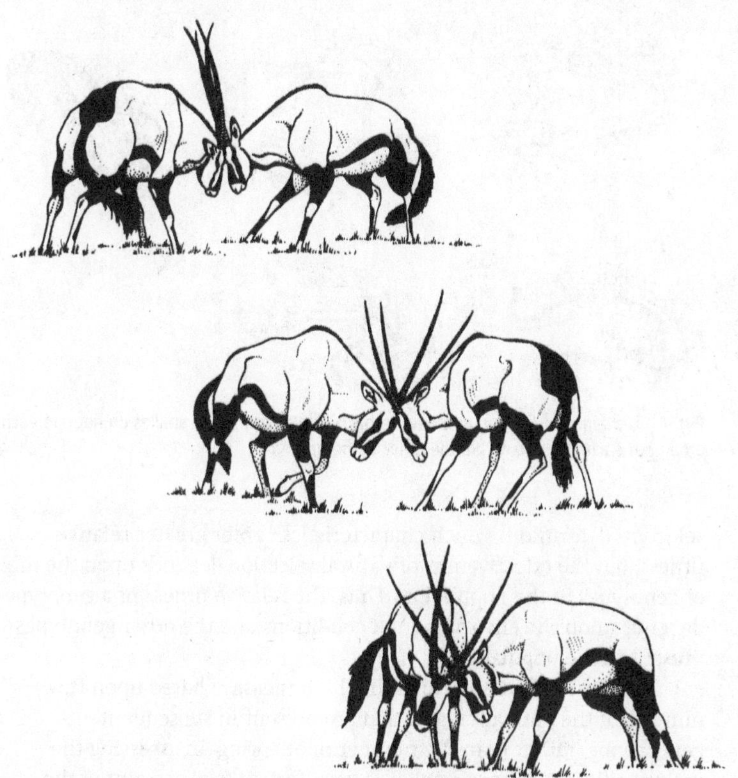

Fig. 3. Ritualized technique of fighting in the oryx (*Oryx gazella*).

fitness A biological and mathematical concept akin to *survival value, which indicates the ability of genetic material to perpetuate itself in the course of *evolution. The concept of fitness may be applied to single genes and to the genetic make-up (genotype) of individual animals, or to animal groups. In animal behaviour, the concept most widely employed is the fitness of a genotype relative to other genotypes in the population. An animal's **individual fitness** is a measure of the relative ability of the animal to leave viable offspring. All factors that affect the animal's fertility and fecundity will affect its individual fitness. These will include the morphological, physiological, and behavioural characteristics of the animal. The process of *natural

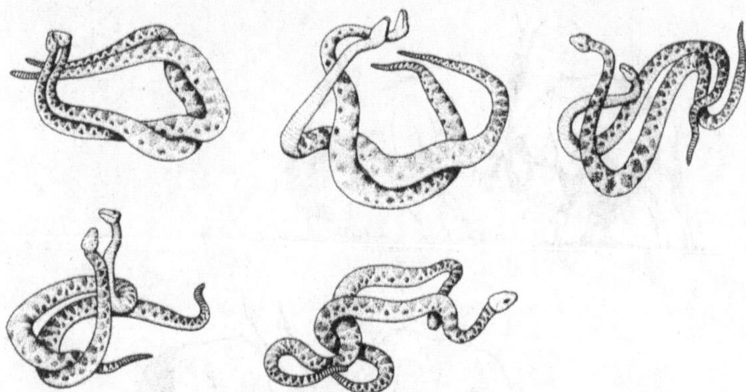

Fig. 4. Technique of fighting in the rattlesnake (*Crotalus* sp.). The snakes do not bite each other, but each attempts to pin the other to the ground.

selection determines which characteristics confer greater relative fitness, but the effectiveness of natural selection depends upon the mix of genotypes in the population. Thus, the relative fitness of a genotype depends upon the environmental conditions and the other genotypes present in the population.

The **inclusive fitness** of an animal is a measure based upon the number of the animal's genes that are present in subsequent generations, rather than the number of offspring. In assessing the inclusive fitness of an animal, it is necessary to take account of the number of its genes that are also present in related individuals. This will depend upon the **coefficient of relationship** between the one individual and another. The coefficient of relationship between parent and child is ½ because a child gains half its genes from each parent. The coefficient of relationship between grandparent and child is ¼, and between siblings is ½. Inclusive fitness is sometimes equated with a simple weighted sum based on the animal's various coefficients of relationship. Thus, it is sometimes seen as the sum of the animal's individual fitness and that of the relatives discounted in proportion to the coefficient of relationship. This measure counts all the animal's offspring, and although this may be a useful measure for practical purposes, it does not accord with the original definition of *Hamilton. Thus inclusive fitness has often been misdefined. It should exclude those of the animal's own offspring that exist because of help received from others, and include those offspring of relatives whose existence is the result of the animal's help being offered to the relative in question.

fixed action patterns Activities that have a relatively fixed pattern of *coordination. They appear to be stereotyped, though they may show variability in *orientation. Many fixed action patterns are controlled without reference to *feedback from the consequences of the behaviour. Examples include the rapid backward escape responses of the squid (*Loligo* sp.), and the song of the cricket (*Gryllus campestris*). Some fixed action patterns are *innate, as are the examples above, while others are the result of *learning. These include many skilled movements, such as the golf-swing, which are ballistic in the sense that, once launched, they are not affected by feedback.

flight Active movement through air. Flight is a form of *locomotion that has evolved only a few times in the history of the animal kingdom. It requires a high degree of specialization of skeletal design, and muscular and nervous physiology, which can only be achieved by animals that are already highly evolved for an agile life on land. Thus flight has evolved among the land-based insects, reptiles, and mammals. The precise course of the evolution of flight among insects is uncertain, because of gaps in the fossil record. By the middle of the Carboniferous era (310 million years ago), there were various distinct orders of flying insects: primitive dragonflies (Anisoptera), cockroaches (*Periplenata* sp.), and grasshoppers (*Trilophidia* sp.). By the Jurassic era flying reptiles had appeared and enjoyed a dramatic radiation before disappearing at the end of the Cretaceous era (70 million years ago). The birds evolved from the same reptilian stock as the dinosaurs and crocodiles. By the Cretaceous era, birds showed most modern skeletal features. Among the mammals, the bats (Chiroptera) are the supreme flyers. They appeared in their present form in the Eocene era (40 million years ago), but little is known about their evolution. Several groups of animals have members that are capable of gliding, but not of sustained flapping flight. These include flying fish (Exocoetidae), gliding frogs (*Rhacophorus* spp.), and lizards (*Draco* spp.). Amongst the mammals there is the gliding possum (*Petaurus* sp.) and the flying squirrel (*Glaucomys* sp.). These animals have skin membranes, stretched between limbs, that enable them to become airborne for short periods.

Flight has a number of functions in the animal's life, of which the most important is *escape from predators. Many flying animals operate close to the limits of muscular and skeletal performance. When flight is used for escape, it is important to have good acceleration. For this reason, both birds and insects are capable of producing brief bursts of muscle power at moments of crisis, and of exploding into flight. Pheasants (*Phasianus colchicus*), for example, are capable of very rapid flight for brief periods only. They use flight only for escape, and otherwise move by walking.

Flight is also used in *predatory behaviour. Birds of prey, such as falcons (*Falco* spp.) and owls (*Tytonidae*), dive on their prey from the air. Insectivorous birds, such as swallows (Hirundinidae), hunt on the wing, as do bats (Chiroptera). Many insects are insectivorous, including dragonflies (Anisoptera) and some wasps (Vespoidea). *territorial and *sexual behaviour also occur during flight in some species. Thus the singing flight of the skylark (*Alauda arvensis*) is a territorial display, as are the spiral duet flights of the speckled wood butterfly (*Paragarge aegeria*).

flocking in birds (For flocking in other animals *see* HERDING.) A flock is a group of birds that remains together as a result of social attraction between individuals. It is in contrast to an *aggregation of individuals that arises when each bird responds independently of the others to some factor in the environment. For example, when a blackbird (*Turdus merula*), a robin (*Erithacus rubecula*), and a hedge sparrow (*Prunela modularis*) are seen together at a garden bird table, they do not constitute a flock, but an aggregation in response to the food at the bird table. On the other hand, a group of wood pigeons (*Columba palumbus*) feeding on a stubble field in winter constitutes a flock. We can recognize that the pigeons are in a flock, and not an aggregation, because they fly from one field to another as a cohesive group.

Flocking often exhibits seasonal variation, being common outside the breeding season. Thus when great tits (*Parus major*) are just beginning to leave the winter flocks and set up *territories, birds may be quite strongly territorial on a warm, sunny day, and revert to flocking if the weather turns cold.

Flocking birds show *coordination of behaviour, both in flight (integrated flight movements), and in behaviour patterns such as feeding, preening, and sleep, which tend to be synchronized within a flock. Such *social facilitation enables animals to exploit opportunities. Other advantages of flocking include increased vigilance, and *dilution of the effect of attacks by predators. Flocks are often formed during *migration, and it has been suggested that this may mitigate errors of *navigation.

food begging Behaviour that elicits donation of food by another member of the species, usually a parent or mate. Juvenile food begging ranges from *vocalization by helpless young to active pestering by relatively grown-up animals. Thus rat pups (*Rattus norvegicus*) suckle at their mother's teats and are initially capable only of simple orientation towards the nipple. Later in development they demand food by specific vocalization and food-begging behaviour. The chicks of herring gulls (*Larus argentatus*) can forage from hatching, but in reality are entirely dependent upon the parents for food. The chick pecks at the red spot at the tip of the parent's bill, to stimulate the parent to regurgitate part of its food. Even fully grown young, quite capable of foraging for themselves, may solicit food in this way.

Juvenile food begging often involves specific *sign stimuli to which the parent responds. For example, nestling blackbirds (*Turdus merula*) stretch their necks and gape in response to movement of the nest, and to the sight of rounded objects in the region of the nest. Inside the gaping mouth of the nestling are distinctive markings that serve as sign stimuli, and induce the parent to deposit food into the gape. *Mimicry of these markings occurs in some brood parasites.

Food begging by adults may occur during *courtship. The female may imitate juvenile food begging to solicit *courtship feeding by her mate.

food chain An aspect of *community life, in which each species occupies a particular *niche that includes food from species further down the chain (*see* Figure 5). Thus each species is an essential component of the community, necessary for the survival of the other species. For example, in the Namib desert community, the climate is so dry that no plants grow, and the terrain consists largely of sand dunes. The prevailing winds carry plant detritus into the dunes, where it accumulates at the bottom of leeward slip-faces. This material consists of fine grass stems and seeds, and provides food for a number of species of beetle (especially *Lepidochora argentorisea*). These animals obtain water from the occasional sea fogs that invade the dunes. The beetles are preyed upon by the white lady spider (*Carparachne alba*) and by the

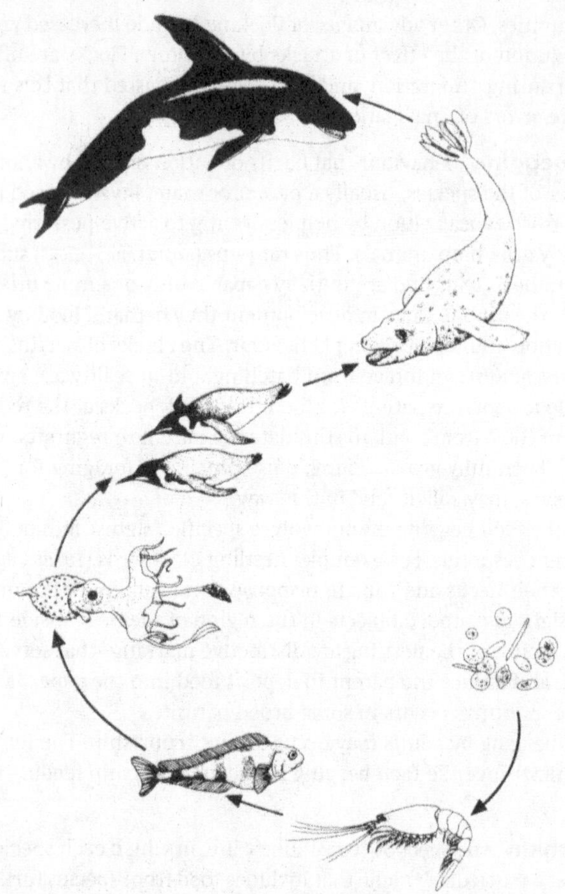

Fig. 5. An example of a food chain, showing the killer whale (*Orcinus orca*), leopard seal (*Hydrurga leptonyx*), king penguins (*Aptenodytes patagonica*), squid (*Histioteuthis* sp.), Antarctic blennies (*Notothenioidei* sp.), krill (*Euphausia* sp.), and plankton.

Namib desert lizard (*Aporosaura anchietae*). The lizard also eats the spider, various flying insects, and wind-blown grass seeds. In turn, it is preyed upon by the dwarf puff adder (*Bitis peringueyi*) that lies in wait under the sand. Other predators include birds of prey that fly in from outside the desert region.

food selection The ability to select beneficial food. Many animals are able to select the right kind of food as a result of their *innate preferences. Others are able to learn rapidly what kinds of food are good for them and what bad.

Feeding specialists consume only one or a few kinds of food. The monarch butterfly (*Danaus plexippus*) eats only milkweed. The koala bear (*Phascolarctos cinereus*) eats eucalyptus leaves. As their food is homogeneous, such animals require a single *hunger detection system, and a simple food-recognition system. Feeding generalists, on the other hand, have a heterogeneous diet, and must distinguish among a variety of *specific hungers or nutritional deficits, and require a complex food recognition system. *Poison avoidance is particularly important for omnivores.

Omnivores can easily become deficient in certain nutrients, such as vitamins and minerals. They may consume too much of a particular food, so obtaining an imbalance of nutrients, and they are in danger of consuming poisonous substances. Omnivores usually rely on *learning to select a balanced diet and avoid harmful substances. Rats (*Rattus norvegicus*), for example, sample novel food substances with caution, eating only a little. If the rat becomes ill after sampling a new food, then it quickly learns to avoid that food in future. This type of learning is an important feature of poison avoidance. Moreover, the rat can learn rapidly which foods are nutritionally beneficial. When a rat suffers a deficit of some essential vitamin or mineral, it takes more interest in novel food substances and samples them more than it otherwise would. If, as a result, it eats food containing vitamins or minerals capable of rectifying its deficiency, then it quickly learns to eat more of that food. Usually, the rat is unable to detect, by smell or taste, the presence of vitamins and minerals in the food. It learns to select the beneficial food on the basis of the post-ingestive consequences of eating it. In other words, the rat can learn to select those foods that make it feel better, and avoid those that make it feel ill.

Many species of omnivorous mammals and birds have been shown to have the ability to learn rapidly to avoid noxious foods and select beneficial foods. Amongst mammals, the ability to recognize different foods is based primarily on *olfaction and *taste, whereas in birds it is based primarily on *vision. However, the principle of food selection based upon post-ingestive consequences remains the same.

foraging behaviour Behaviour associated with searching for, subduing, capturing, and consuming food. *Searching may take a variety of forms, including active *hunting, searching by browsing and grazing animals, and (purely sensory) searching by sit-and-wait predators. The **tactics** of a forager are the methods by which it goes about the business of obtaining food. For example, the hunting groups of spotted hyena (*Crocuta crocuta*) adopt the tactic of running after their prey and overtaking them by superior stamina and by cooperative hunting manoeuvres. The solitary cheetah (*Acinonyx jubatus*), on the other hand, stalks its victim and then overcomes it in a sudden high-speed dash.

Foraging **strategy** is a type of *evolutionary strategy concerned with the costs and benefits of foraging. Natural selection favours efficient foraging, and most animals are extremely adept at searching for, and harvesting, food. Foraging efficiency is usually a matter of **trade-off** between competing priorities. These may include energy gained versus energy spent; energy gained versus risk of predation; energy gained versus losses to rivals, etc. Such trade-offs apply not only to energy produced by foraging, but also to that usurped. Thus the food-finding abilities of one species can become part of the foraging strategy of another species. In other words, the foraging investment of one species (the producer) is parasitized by another species (the scrounger). For example, lapwings (*Vanellus vanellus*) foraging for earthworms (*Lumbricus*) are pirated by black-headed gulls (*Larus ridibundus*). Often the gulls are involved in expensive aerial chases. They can avoid these if they attack the lapwings just as they obtain their prey. While taking large worms, the lapwings adopt a tell-tale crouching posture, which the gulls use as a cue as to the best time to attack. Lapwings can reduce their vulnerability to attack by concentrating on the smaller worms nearer the surface, and thus obviating the crouching required to obtain the larger, deeper, worms that are the gulls' principal target.

Animals must expend energy in order to forage, and there may be circumstances in which the energy available to spend is limited. Similarly, there may be a limited amount of time available for foraging, and the rate at which a forager can harvest prey may also be constrained. In particular, the **handling time** required to recognize, capture, and process each food item may place a limit on the harvesting rate. Thus the foraging animal has to operate within its energy and time *budget.

The **profitability** of a food item is the net energy value divided by the handling time. When an animal is able to choose among prey items, we might expect it to choose the most profitable prey. For example, bluegill sunfish (*Lepomis macrochirus*) feed on water fleas (*Daphnia* spp.). When these are scarce, the fish show no size preference, but when they are abundant the fish take only the largest and most profitable.

Animals usually consume their food on the spot, but if an animal has a nest, or comes from a colony, it may take food home. Such animals usually make an outward journey, spend some time searching, and then make a return journey. This type of foraging is called **central place foraging**. The distance between the central place and the foraging site is usually called the **travel distance**. In general, provided food availability remains constant, the optimal load carried by the animal increases with the travel distance. Time spent at the foraging site also increases with the travel distance. The return travel is usually shorter than the outward travel, because the animal weighs more on its return trip and should minimize the time spent carrying the load. An example can be seen in desert ants (*Cataglyphis fortis*) that have meandering outward foraging paths and straight return paths. The central place for a group of foraging animals can act as an **information centre**. This phenomenon is found in foraging trail-laying ants (Hymenoptera), honey-bees (*Apis Mellifera*), and some colonially nesting birds.

forgetting *See* Memory.

Frisch, Karl Ritter von **(1886–1983)** Eminent ethologist. He was born in Vienna and educated in Vienna and Munich. His academic career started in 1910, when he became an assistant at the Zoological Institute at Munich. He subsequently held chairs at Rostock University (1921), Breslau (1923), and Munich (1925). In 1973 he shared the Nobel Prize for Medicine, for his contributions to the study of animal behaviour. His most famous research is his discovery that honey-bees (*Apis mellifera*) inform each other of the whereabouts of food, by means of a specific dance.

frustration A state of *motivation that arises in situations in which the consequences of behaviour are less than those that the animal has been led to expect on the basis of past experience. For example, a hungry animal is likely to become frustrated if it is physically prevented from obtaining food that it can see, if an expected food reward is delayed, or if its food is less palatable than usual.

Frustration involves an element of expectancy, whereas deprivation results in a desire or need, but without any associated expectation of attaining the desired state. In a totally strange environment, devoid of any recognizable external stimuli, an animal would be deprived of food, affection, etc., but not frustrated. An animal in a familiar environment, where there are stimuli associated with food, etc., and stimuli associated with special times of day, such as meal times, would have expectations based upon past experience with those stimuli, and would become frustrated if food did not become available.

The behaviour seen during frustration includes *conflict behaviour, and *displacement activity. Prolonged frustration may lead to *stress.

function (of a trait) The increase in *reproductive success that the trait confers on its possessor's *genetic constitution, or **genotype**. For example, the function of *nest-building is to provide a nest to keep the occupants warm, and to help protect them from predation. This implies that these consequences of nest-building confer higher reproductive success on the genotype of the individuals that practise it.

In most cases, traits that increase the reproductive success of the genotype that produces them also increase the success of their individual carriers, but this is not always true. For example, genes that cause their carriers to assist other individuals sharing the same genotype (i.e. relatives) may be favoured by *natural selection and may spread through the population, even if they lower the reproductive success of their carriers. This phenomenon is called *kin selection.

Functions are always complex and usually multiple. For example, the function of *incubation in herring gulls (*Larus argentatus*) is dependent upon a number of consequences of incubation. Thus, incubation keeps the eggs warm, but it prevents the sitter from foraging for food. It protects the eggs from predation, but it exposes the adult to attack by larger predators. Thus incubation has both costs and benefits. The function of incubation, therefore, has to be seen in terms of all the consequences of incubation that affect reproductive success, in comparison with the consequences of not incubating, or of incubating in a different manner.

functional reference A form of communication in which the signal varies according to the type of object to which reference is being made. It is most commonly seen in *alarm situations. A number of species, including chickens (*Gallus gallus domesticus*) and various types of monkey (Simiae), have been reported to give alarm calls that differ according to the type of danger. Adult vervet monkeys (*Cercopithecus aethiops*) give different alarm calls when they sight a leopard (*Panthera pardus*), a python (*Pythonidae*), or an eagle (Accipitridae). When they hear the alarm, the other monkeys take action appropriate to the predator that has been sighted. If it is a snake, they look down, and if it is an eagle they look up. If they hear the leopard alarm, then they run into the trees.

G

game theory A type of mathematical modelling used in decision theory, economics, and evolutionary theory. It is used to analyse *evolutionarily stable strategy, a type of *evolutionary strategy that cannot be bettered by any feasible alternative strategy. Often the best strategy (i.e. evolutionary ploy) for an individual agent (genetically distinct type) depends upon the strategies adopted by other members of the population. In a symmetric game, the two contestants start in identical situations, and have the same choice of strategies and the same prospective pay-offs. There may be differences between them, but if these are not known to the contestants, they cannot affect their choice of strategies.

In an *asymmetric game there is usually a perceived difference between the contestants, which affects their choice of strategies. The contestants may differ in sex, size, age, or a resource such as territory. For example, an individual could behave like a hawk when it is the owner of a territory and like a dove when it is an intruder into the territory of another. That is, when holding a territory, it fights to kill or injure the opponent, even though there is a risk of injury to itself. When an intruder, it threatens its opponent but avoids serious fighting. This is called a bourgeois strategy. (For a simple example, *see* ASYMMETRIC GAMES.)

generalization A response made to a stimulus which is similar to the **conditional** stimulus. When an animal has learned a particular response to a particular stimulus, it may also show the response to other similar stimuli. This phenomenon, known as stimulus generalization, was first described by *Pavlov.

The normal explanation of stimulus generalization is called the common elements theory. A stimulus used in a *conditioning experiment consists of a collection of separate elements. For example, a tone has a given frequency, intensity, and duration. These dimensions of the stimulus may become conditioned during acquisition of a conditional response. A new stimulus that has some elements in common with the conditional stimulus may be capable of eliciting the conditional response to some extent. Whereas humans distinguish readily between a 1000 Hz tone and a 300 Hz tone, pigeons (*Columbia livia*) treat them as similar. The tones have other qualities in common, especially the fact that they are unlike natural sounds in being characterized by a single frequency. It is perhaps not surprising, therefore, that pigeons treat them as similar. If a pigeon is rewarded in the presence of a 300 Hz tone then it has a lesser tendency to respond to tones of other frequencies. *Discrimination training reduces the number of elements that are associated consistently with reward. The pigeon is encouraged to pay *attention to the frequency of the tone by the fact that the other dimensions of the sound are not associated consistently with reward.

genetic relatedness (r = coefficient of) A measure of the probability that a gene in one individual will be identical by descent to a gene in a particular relative. An equivalent and alternative measure is the proportion of an individual's genome that is identical by descent with the relative's genome. Note that the genes must be identical by descent and not simply genes shared by the population as a whole. This means that r is the probability that the gene common to two individuals is descended from the same ancestral gene in a recent common relative.

g

genetics of behaviour The study of inherited characteristics that affect behaviour. This can be achieved by a number of methods, including breeding studies, inbreeding studies, strain and race comparisons, segregation analysis, diallele crosses, and DNA studies.

In breeding studies, the investigator controls the system of mating, and selects those animals to be mated on the basis of the presence or absence of some particular phenotypic trait. This type of artificial selection has been used for centuries for the improvement of domestic animals. When a trait is controlled by a single gene with only two alleles, selection for the recessive allele can be completed within one generation. Selection for the dominant allele takes longer. In the fruit-fly (*Drosophila melanogaster*) there is a sex-linked mutant yellow gene. Males with this gene are less successful in mating with wild-type females than are wild-type males. The yellow males have an altered courtship pattern that reduces their mating success. They are less stimulating to the females because their courtship contains a smaller proportion of wing vibration.

Inbreeding is a special form of artificial selection in which closely related individuals are mated. It provides a means of increasing the genetic homogeneity of a population. Different inbred lines known to be genetically unalike can be compared and analysed for phenotypic differences. Comparisons of inbred strains subjected to a range of environmental treatments yield data on heredity–environment interactions. For example, rats (*Rattus norvegicus*) can be selectively bred for their ability to solve maze puzzles. When such animals are raised in an impoverished environment, there is little difference between the maze-bright and maze-dull strains. Both perform badly. When raised in a normal environment, the maze-bright strain performs well and the maze-dull strain performs badly. When raised in an enriched environment, both strains perform well.

Many of the data that have been gathered in behaviour–genetic studies come from comparisons of inbred strains of different races occurring in nature. For example, different breeds of dogs (Caninae) differ in behaviour as well as physical characteristics. These differences can be studied by cross-breeding experiments, the results of which are compared with what would be expected under various genetic hypotheses.

The *heritability of behaviour is measured in terms of the variance in phenotypic traits within a particular population. The heritability of a trait is the fraction of the observed variance that is due to differences in heredity. It can be studied by looking at differences in closely related animals raised in different environments, and at differences among animals raised in the same environment. In recent years it has been possible to compare the DNA of different specimens.

geotaxis An aspect of *orientation in which the animal performs a *taxis in relation to gravity. Many animals possess *sense organs capable of detecting the direction of gravity. Thus jellyfish (*Medusae*) have statocysts, each being a cavity that contains a small pebble-like object, the statolith, the position of which can be detected by *mechanoreceptors in the wall of the cavity. A similar arrangement is found in the vestibular apparatus in the ear of vertebrates. Geotaxis consists simply of moving directly towards, or away from, the direction of gravity.

goal The state of affairs that brings a particular activity to an end. This may be internal to the animal, such as *consummatory behaviour that brings to an end a period of *appetitive behaviour. In some cases, simply performing a consummatory act is sufficient to terminate a particular activity. It may also be an external state of affairs, such as a source of food. Upon reaching the goal, appetitive behaviour normally ceases. Thus male copulatory behaviour ceases after ejaculation (the consummatory act), and foraging behaviour ceases when food is found, and is followed by eating (the consummatory act).

It is helpful to distinguish among goal-achieving, goal-seeking, and goal-directed behaviour. A **goal-achieving** system is one that can recognize the goal once it is arrived at, but the process of arriving at the goal is largely determined by environmental circumstances. A **goal-seeking** system is one that is designed to seek the goal without the goal being represented explicitly within the system. Many physical systems are of this type. A **goal-directed** system involves an explicit representation of the goal-to-be-achieved, which is instrumental in directing the behaviour. In animal behaviour it is sometimes postulated that the animal possesses a **sollwert* or *search image. Such an animal would be capable of goal-directed behaviour. The various types of goal-oriented behaviour are important in accounting for the apparently *purposive and *intentional behaviour of animals.

gravity receptors An arrangement of *mechanoreceptors that detects the direction of gravity. Invertebrates typically have some kind of **statocyst**, a cavity that contains a **statolith**, the position of which in the cavity can be detected by mechanoreceptors in the wall of the cavity. In many marine animals the statolith is a grain of sand that is incorporated into the statocyst. If lobsters (*Decapoda*) are kept in an aquarium containing iron filings, then these get taken up into the statocysts. By using a magnet, it is possible to alter the apparent direction of gravity, and the lobsters change their *orientation accordingly. In vertebrates a similar principle operates in the **vestibular apparatus** of the inner ear.

gregariousness The tendency of animals to form social groups, as in the *schooling of fish, the *flocking of birds, and the *herding of mammals. Such groups should be distinguished from *aggregations, which result from attraction to an environmental feature, such as a food source. Gregariousness results from mutual attraction and usually involves distinct *social relationships. Often there is a *social organisation, which is attuned to the animals ecological *niche.

grooming All forms of care and attention to the body surface, either by an individual or by conspecifics. Maintenance of the body surface, whether it be the exoskeleton of insects, the scales of fish and reptiles, the plumage of birds, or the pelage (fur) of mammals, is necessary for protection against attack from environmental factors, and such biological factors as parasites and micro-organisms. Grooming also serves to remove surface debris from peripheral sense organs, and sometimes to apply waxy or oily secretions to maintain the water-resistant properties of the body surface.

Animals without limbs or other appendages, such as fish, can groom only by rubbing themselves against an object in the environment. Some also enter into a *symbiosis with specialized cleaner fish. Animals with limbs use them to scratch, remove debris, and apply secretions. They may also use their mouth, tongue, teeth, and tail. Autogrooming, where the animal grooms itself, is a routine matter in most species.

Allogrooming, where one individual grooms another, is a common aspect of social behaviour, especially between parents and offspring and mated pairs. It is often directed to regions that the recipient cannot itself reach or inspect. Amongst highly social animals there is often a grooming *hierarchy, in which the subordinate grooms the dominant individual. Social grooming is thought to strengthen *social relationships and may serve as *appeasement.

group living The result of *gregariousness. The main factors affecting group living are food and predation. *Foraging may be enhanced if members of a group inform each other about the whereabouts of food, and may be rendered more efficient by *cooperative behaviour. On the other hand, if food availability is restricted, then there may be competition for food, if the group it too large, and there may be a *disturbance effect during foraging. Thus redshank (*Tringa totanus*) are more likely to disturb each other's shrimp (*Corophium* spp.) prey, if they feed in a large group. This happens because the birds' footsteps trigger the shrimps' retreat behaviour.

A specific advantage of living in a group is increased *vigilance against predators. Many predators rely on an element of surprise to catch their prey, so constant vigilance is a good insurance against predation. Feeding efficiency is reduced if the individual spends too much time watching out for predators, but an individual in a group can, to some extent, rely on the vigilance of others. Another advantage of group living is the *dilution effect. By joining a group, an individual dilutes the effect of a successful attack by a predator, because there is a chance (depending upon group size) that another individual may be the victim. Horses (Equidae) in the Camargue are less likely to be attacked by flies, if they are in a large group. Disadvantages of group living may include *competition for mates and increased *parasitism.

group selection A theory that claims that natural selection will favour behaviour that reduces the fitness of the donor if it benefits the group or species as a whole. However, it is difficult to demonstrate how such a situation could have arisen during evolution.

The main problem is to prevent cheating. For example, let us imagine a population of rabbits (*Oryctolagus cuniculus*), in which members do not warn each other of approaching danger. Suppose that a gene is introduced that promotes thumping on the ground by a rabbit that senses danger. The thumping serves to alert other rabbits, but it also attracts the attention of the predator. Thus, it would seem that the neighbouring rabbits benefit from the *warning while the thumper endangers itself. If we assume that the thumper manages to pass on the thumping gene before being eaten by a predator, then we have a subpopulation of thumping rabbits within a population of non-thumpers. Because thumping is of benefit to the group, the group selection argument maintains that the thumping group will evade predation better than the non-thumping group.

This situation is not evolutionary stable, because a rabbit in the thumping group that did not possess the thumping gene would not endanger itself when it detected a predator, yet it would benefit from the warnings given by other members of the group. This rabbit would be a cheat and would have an advantage over other members of the group. The greater reproductive success of the cheating rabbits would mean that the thumping gene would gradually be eliminated from the thumping population.

Warning is a type of *altruism that is usually accounted for in terms of *kin selection, rather than group selection.

gustation A form of chemoreception synonymous with *taste.

habitat The natural home of an animal or plant: the external environment to which it has become adapted during the course of *evolution. Habitats are usually described in terms of salient physical and chemical features of the environment, and are largely determined by climate. Each species has a specific *tolerance range of environmental factors, which depends largely upon their physiological mechanisms and their ability to adjust to environmental changes by means of appropriate behaviour.

Many animals have some ability to select their habitat, and thus increase their chances of survival. In some species this is a matter of simple *orientation. For example, when common woodlice (*Porcellio scaber*) are placed within a gradient of humidity, they move about in an irregular manner. These movements are more rapid in dry than in moist air, with the result that the animals spend more time in damp places, thus achieving a very simple form of habitat selection. More sophisticated animals achieve habitat selection via a mixture of *innate preferences and *imprinting.

habituation An aspect of *learning in which repeated applications of a stimulus result in decreased responsiveness. For example, the escape response of a guppy (*Poecilia reticulata*) to a shadow passed overhead diminishes progressively if the stimulus is presented every 2 min. Eventually the fish does not respond at all.

A common feature of habituation is that the habituated response reappears if the stimulus is withheld for a long period of time. Thus the guppy escape response reappears if no shadows are presented for about a day after the response has habituated. Usually, a response habituated to one stimulus shows *generalization to another similar stimulus. Thus, if a new stimulus is similar to one to which the animal has already habituated, then habituation will be more rapid. The recovery of a response upon presentation of a new stimulus is called **dishabituation**.

Habituation is a widespread phenomenon in the animal kingdom. Its *survival value lies in counterbalancing the advantages and disadvantages of responding to stimuli of uncertain significance. No animal should ignore potentially dangerous stimuli. On the other hand, over-responsiveness is a waste of time and energy.

Hamilton, William Donald (1936–2000) British evolutionary biologist. His main contribution to the study of animal behaviour was his explanation of how natural selection acts on social behaviour.

He was educated at Tonbridge School and Cambridge University. He was a lecturer at Imperial College, London from 1964–77, a professor at Michigan University from 1978–84, and then a Royal Society Professor at Oxford University until his death. He received many international scientific prizes, and is widely regarded as the most important Darwinist since Charles Darwin.

handling time *See* Foraging behaviour.

hatching The process by which a juvenile animal emerges from the egg outside the body of the parent. In many animals, including invertebrates, fish, and amphibians, hatching is facilitated by the secretion of enzymes that digest the egg envelope or membrane. Reptiles, birds, and monotremes have tougher eggshells, and hatching is largely the result of vigorous movements of the embryo. Many reptile and bird embryos develop an egg-tooth that is used to tear an opening in the shell prior to hatching. Echidnas (*Tachyglossidae*) (monotreme mammals) also have an egg-tooth used for this purpose.

hearing The detection and perception of sound. The complexity of the auditory apparatus ranges from the two-celled units of moths to the sophisticated hearing system involved in bat (Chiroptera) *echolocation.

Sound is made up of rapid, small changes in pressure within the surrounding medium (e.g. air or water), that originate from a vibrating source and propagate outwards in waves. The intensity of sound is expressed on a logarithmic scale as decibels of sound pressure level (dB SPL). The logarithmic scale makes it possible to cover the great range of pressures involved in hearing. The range of human hearing is 120 dB. The frequency of sound is the number of complete pressure cycles occurring in one second. The unit of frequency is called the Hertz (Hz; 1 cycle/s) or the kilohertz (kHz; 1000 cycles/s).

The ears of animals have two general features. A peripheral device for converting sound pressure to vibratory motion, and a specialized sense organ for converting this motion into nerve impulses. The true substrate of hearing lies in the coded nerve messages that are processed in the central nervous system. Moths (Lepidoptera) have the simplest known peripheral auditory system. They have two ears, which allows for directional hearing. Each is composed of a **tympanic membrane** and two acoustic receptor (nerve) cells embedded in a strand of tissue attached to the membrane. Moths are able to discriminate between faint and intense sounds over a wide range of frequencies, but they are tone deaf, the brain not making use of all this information. They are capable of responding to rapid temporal changes, such as the very brief pulses of sound in the cry of a bat. They can determine the distance and direction of the sound, and if the bat is far away they fly directly away from the source. If, however, the bat cries are close and intense, the moth takes evasive action, dropping with wings folded, or going into a power-dive.

Vertebrates usually have an outer, a middle, and an inner ear. The outer ear, especially in mammals, serves to collect and funnel sounds to the middle ear. The middle ear is composed of a membrane, the tympanum or eardrum, and, in all but the simplest systems, one to three ossicles (bones) that conduct the vibrations of the tympanum to the inner ear. The inner ear is a fluid-filled bony capsule containing an array of numerous receptor cells. Frogs (Anura) have a frequency range between 1.0 and 300 kHz and their hearing is tuned to the mating call that is characteristic of their species. The auditory neurons of one species do not respond to, nor can they be stimulated by, the calls of a different species, even though they may share the same habitat. Thus the hearing systems of frogs act as an important *isolating mechanism.

Birds and mammals have auditory capabilities far beyond those of other vertebrates, thanks largely to their larger brains and greater processing power. Owls (Strigiformes) and bats (Chiroptera), have particularly advanced auditory location capabilities. Owls are able to hunt at night, not only because of their keen nocturnal *vision, but also because they can determine accurately both the direction and distance of the source of sounds generated by prey, even in the pitch dark when vision is impossible. This ability stems from the asymmetry in the size and placement of their external ears, which endows the owls with a passive sonar system. Bats have an active sonar system, in which they themselves produce ultrasonic cries and subsequently monitor the time delays and intensity changes in the reflected sound waves. This *echolocation system enables them to catch moving prey at night.

herding A form of *social organization in mammals. Like *schooling in fish and *flocking in birds, herding confers advantages primarily in connection with predators. Members of a herd benefit from the *vigilance of their companions, in the sense that many pairs of eyes are better than one. This is especially true when the mode of feeding is such that detection of predators is difficult without interrupting feeding. A grazing animal has to look up periodically to scan the environment for possible danger, but if the animal is feeding in a herd it does not have to do this so often, and can thus spend more time feeding.

Members of a herd sometimes cooperate to deter predators. In response to attacks by wolves (*Canis lupus*), musk oxen (*Ovibos moschatus*) gather into a defensive formation that presents an array of horned heads to the predators and protects the more vulnerable animals behind them. The larger the herd, the lower the probability that a given individual will be the victim of an attack. Carnivore predators usually single out a young, old, or infirm individual, thus a healthy animal benefits greatly from being a member of a herd.

The structure of a herd is intimately related to the ecology of the animal. Small groups of carnivores or primates, which have some *cooperative behaviour, have social relationships that tend to be long lasting. A consequence of small, stable groups is that there is a tendency toward inbreeding. However, in most species, *incest is not common, because young animals tend to leave their natal group and join or found another. Among lions (*Panthera leo*), hyenas (*Hyaena*), baboons (*Papio* spp.), and most other primates, it is the males that leave. Among wild dogs (*Lycaon pictus*) and chimpanzees (*Pan troglodytes*), it is always the females.

*Territory formation often occurs in large groups of antelope (Bovidae), even though they may be continually on the move. Ugandan kob (*Adenota kob thomasi*) herds are often made up of distinct group as well as some solitary animals. The groups may be made up of males of assorted ages, or of females and young, escorted by one or two mature males. The single animals are the territorial males, each holding an area of about 500–900 m^2. The territories are defended against rival males, usually would-be territory holders from the bachelor groups. Female kob are attracted by the *display of the territory holder, and after mating takes place, she rejoins her group. Thus the male territories make up an arena, or *lek, a form of social organization more commonly found among birds.

Herd organization varies considerably from species to species, but is usually has two aspects: control of females by competing males, and protection of the young. The former often leads to synchronized breeding, which leads to so many juveniles being produced in such a short time that predators cannot possibly kill them all.

heritability of behaviour The fraction of the variance observed among members of a population that is due to differences in heredity. When we examine a particular trait, such as body weight, in a population of animals, we are looking at a phenotypic value that is made up of both genotypic and environmental components. By looking at related populations in different circumstances, it is sometimes possible to estimate how much of the variability between individuals is due to environmental factors, and how much to genetic factors.

There are various statistical methods of estimating the heritability of behaviour. Some are based upon examining phenotypic variation in genetically identical individuals. The total phenotypic variability among genetically identical individuals is compared with the total phenotypic variability in a natural genetically variable population. The comparison generates a ratio which contrasts the genetic versus environmental components of phenotypic variability.

Estimates of heritability may be subject to various types of error. Interaction between genotype and environment may introduce variability that is not taken account of in ordinary heritability calculations. For example, rats (*Rattus norvegicus*) can be selectively bred to be good (maze-bright) or poor (maze-dull) at running in a maze. However, if maze-dull rats are raised in a restricted environment, they perform poorly, but if they are raised in an enriched environment, they perform well in a maze. The same is true of maze-bright rats. The difference between maze-bright and maze-dull rats only shows up when both are raised in a normal environment. Correlation between genotype and environment, another source of error, can arise when individuals select particular environments, or develop particular habits, to compensate for genetic defects.

Another source of possible error is non-random mating. This is particularly true of humans, for certain traits, such as height and intelligence. People tend to choose partners of similar height and intelligence quotient (IQ). Non-random mating is likely to occur with respect to any traits that are subject to *sexual selection.

heterogeneous summation Additivity of independent and heterogeneous features of a stimulus situation in their effects upon behaviour. For example, the herring gull (*Larus argentatus*) recognizes its eggs by various features, including size, shape, background colour, and speckling. These features have been shown to be additive in their effects upon retrieval of eggs that have rolled out of the nest. Heterogeneous summation is not a universal law of visual recognition, but it has been shown to hold in a number of cases.

hibernation A form of winter *dormancy that is characterized by slowing of metabolic processes and a marked fall in body temperature. True hibernation should be distinguished from other types of dormancy. True hibernators have *thermoregulation that allows the body temperature to fall to that of the surrounding air, maybe as low as 2°C. True hibernation is found in some mammals and birds.

Among the mammals, hibernation occurs in monotremes, marsupials, rodents, insectivores, and bats (Chiroptera). True hibernation is found only in the smaller species, although other forms of dormancy occur in other species, such as bears (*Ursus* spp.). Because of their proportionately large surface, small animals cool more readily than large ones. They also warm up more quickly, due to their small thermal capacity.

Many birds can enter a state of torpor in which their body temperature falls to low levels. The white-throated poorwill (*Phalaenoptilus nuttallii*) and nightjar (*Caprimulgus europaeus*) can endure body temperatures as low as 6°C without ill effect. Torpor usually lasts only a few hours in birds, and appears to be confined to the inactive phase of the daily cycle. Seasonal hibernation is known in only a single bird family, the goatsuckers (*Caprimulgidae*).

Hibernation is characterized by a state, similar to *sleep, in which the rate of heartbeat is lowered and breathing is slowed. The body temperature is lowered and energy expenditure is greatly decreased. Hibernating animals often adopt a sleeping posture, and choose a typical sleep-site in which to hibernate. Some, such as the ground squirrel (*Citellus lateralis*) accumulate body fat prior to hibernation, while others, such as the golden hamster (*Mesocricetus auratus*) do not accumulate fat, but store food and build a nest. Many hibernating animals awake periodically, some to eat and drink.

Some hibernators, such as the ground squirrel, are known to have marked *circannual rhythms underlying their seasonal hibernation. Under natural conditions the ground squirrel hibernates for a three- or four-month period, during which it progressively loses weight. After hibernation, body weight increases up to the onset of the next hibernation in October. When isolated in the laboratory, at a constant temperature, the rhythm of hibernation and associated weight changes may persist for up to two years. If serum from a hibernating ground squirrel is injected into a non-hibernating ground squirrel, then hibernation is induced. It appears that some basic physiological mechanism is responsible for controlling hibernation, and that this is driven by a *biological clock, more or less independently of the prevailing external conditions. In other hibernators, a fall in environmental temperatures appears to trigger hibernation, but hibernation appears to be little affected by level of illumination or day length.

hierarchy A principle of organization that occurs at many levels in the control of behaviour. In a hierarchy, the elements are ordered in such a way that higher-ranking elements control lower ones. This is true of the organization of muscles, and of motor control in general. The organization is hierarchical in the sense that the detailed instructions are not all formulated at the highest level, but at each level commands are issued on the basis of more general instructions from the level above.

Hierarchical principles have often been invoked to account for large portions of the animal's behavioural repertoire. For example, we might describe the *nest-building behaviour of a bird in terms of a hierarchy of *goals. The overall goal of owning a nest may generate a goal of gathering twigs and another goal of fastening twigs into the nest. The process of fastening may itself be broken down into various sub-goals. Although hierarchies are regarded an important principle of behavioural organization, there is little agreement as to their precise role.

Hierarchies also occur in the *social organization of behaviour. *Dominance hierarchies are found amongst farmyard chickens (*Gallus gallus domesticus*), where individual A dominates B, and B dominates C, etc. Such linear hierarchies mean that there are never members of equal rank as there would be with a branching hierarchy. Amongst highly social animals there is often a grooming hierarchy, in which the subordinate grooms the dominant individual. Social grooming is thought to strengthen *social relationships and may serve as appeasement.

hoarding The storage of food, or other things, either in a central cache, or distributed throughout the *home range or *territory. Storing food in a single, large cache, sometimes called **larder hoarding**, occurs in many small mammals, and in some birds and insects. Acorn woodpeckers (*Melanerpes formicivorus*), for example, live in small groups and prepare trees by drilling hundreds of evenly spaced holes in tree trunks and branches. Each hole is filled with an acorn. Bumble bees (*Bombus* and other genera) construct separate vessels in their nests for the storage of pollen and nectar.

Scatter hoarding, the storage of food in many dispersed sites, is common in many birds and mammals. The grey squirrel (*Sciurus carolinensis*) stores food in holes that it digs in the ground, which are carefully covered with leaves and grass. Crows (*Corvus*), jays, magpies (*Pica*), and nutcrackers (*Nucifrag*) do the same.

Food hoarding serves many functions. Periods of low food availability can be survived by eating food stored during periods of abundance. Jays (*Garrulus glandarius*) rear their young on stored food that would otherwise be unavailable when the young are in the nest. Carnivores, such as leopards (*Panthera pardus*), and weasels (*Mustela* spp.), cache prey to prevent its loss to scavengers.

Animals that do not hoard food often exploit food stored by others. Willow tits (*Parus montanus*) follow food-hoarding coal tits (*Parus ater*) and rob their caches. Scatter hoarders and larder hoarders deal with this type of *parasitism in different ways. Larder hoarders vigorously defend their caches. Scatter hoarders cannot do this, and instead they reduce the likelihood of losing their caches by spacing them out and using a variety of types of place for storage. Scatter hoarding makes it uneconomical for other animals to exploit the caches, but it also presents the hoarding animal with the problem of recovering its own stored food. Some species are known to remember accurately the location of their many caches. The red fox (*Vulpes vulpes*), the common jay, and the marsh tit (*Parus palustris*) all remember the location of hoarded food and readily find it again.

homeostasis Regulation of the internal state of the body. Animals can be divided roughly into **conformers**, which allow their internal environment to be influenced by external factors, such as temperature, and **regulators**, which maintain their internal environment in a state that is largely independent of external conditions. The processes by which regulators control their internal state come under the general heading of homeostasis.

For example, when human body temperature rises above 37°C, cooling mechanisms, such as flushing and sweating, are brought into action. When temperature falls below the optimal level, warming mechanisms, such as shivering, come into play. By employing a number of such finely tuned **feedback** mechanisms, humans are able to achieve a precise *thermoregulation and thermal homeostasis.

Homeostasis is achieved not only by regulatory processes involving *feedback, but also by **anticipatory** mechanisms. For example, in the rat (*Rattus norvegicus*) and pigeon (*Columbia livia*), drinking occurs as a direct response to environmental temperature change, in anticipation of any change in fluid balance arising from thermoregulation. In many animals, cooling involves loss of water in panting or sweating. Such water losses would make the animal thirsty, if the animal had not already forestalled this effect by drinking prior to the loss of water.

In addition to purely physiological responses, homeostasis may also involve behaviour. Thus an animal may seek shade to avoid overheating, or it may immerse its body in water. A thirsty animal will generally eat less (to cut down on the water lost in excretion) and seek cool places (to cut down on water lost in thermoregulation).

home range The area in which an animal normally lives, regardless of whether or not the area is defended as a *territory, and without reference to the home ranges of other animals. Home range does not usually include areas through which *migration or *dispersion occur.

homing Returning home after an absence, whether it be a *foraging trip or a *migration. Adelie penguins (*Pygoscelis adeliae*) may leave their young for up to 2 weeks while they forage far out at sea. They are able to return to their nest even if displaced by storms.

To be able to return home successfully, from a considerable distance, requires a considerable feat of *navigation. The homing animal not only has to set off in the right direction, but also has to maintain the correct course despite changing weather conditions. A number of navigational cues are known to be used by homing animals, although their mode of deployment is not fully understood.

An important factor is time. Apparent time changes with one's position on the Earth, and an indication of position can sometimes be gained by comparing the apparent local time with one's personal time. Many animals are able to use their internal *biological clock to keep track of personal time. The internal clock will be synchronized to the apparent time at home. If the displaced animal's internal clock indicates that it is ahead of the local apparent time, then the animal must be west of home; if behind, it must be east of home. Many birds are known to use their internal clocks in this way.

Other navigational cues used by homing animals include changes in the Earth's magnetic field, the pattern of stars at night, the polarization of sunlight, and visual landmarks.

hormones Chemicals (usually peptides or steroids) that are produced by the glands of the *endocrine system of vertebrates, and are also found in invertebrates. Hormones are released into the blood in very small amounts, but most hormone molecules have a life of less than 1 h in the blood, and have to be secreted continuously if their presence is to be effective. There is usually a series of events that provides negative feedback, which regulates the level of hormones in the blood.

Each hormone has specific functions and often affects specific target organs. The control of hormone secretions is achieved either by direct nervous action, or by the action of other hormones. There are three main ways in which hormones influence behaviour: (1) they can influence effectors, such as special structures involved in behaviour (e.g. the comb of a cockerel (*Gallus domesticus*)); (2) they can influence peripheral *sensory receptors and so modify the input to the brain; (3) they can influence the *brain directly. Hormonal influences upon behaviour may be slow and prolonged, or quick and short lived.

Hormones have important effects upon the *ontogeny of *reproductive behaviour, *migration, and other seasonal activities. Many *rhythms, such as the *sexual and *menstrual cycles, are controlled by hormones. Hormones also have a short-term influence upon *parental and *sexual behaviour, as well as aspects of *motivation, such as *fear.

hunger An aspect of *motivation associated with food. Hunger has both *appetitive and *consummatory aspects. However, food gained is not always eaten. It may be lost, hoarded, sold, donated to another, or eaten. Food that is eaten is digested to release energy and nutrients and waste products.

Hunger may result from energy deficit, from a deficit of certain nutrients, from detection of palatable food, from cues (such as time of day) associated with *feeding very quickly to consume this type of food. On the other hand, if ingestion of a novel food is followed by sickness, then the animal may quickly learn never to touch that food again. Experiments have shown that the beneficial, or deleterious, substance does not have to be present in the food. So long as ingestion is followed by the good or bad consequences (e.g. the substance is injected after feeding) then the animal will learn to seek, or avoid, the food in the future. These experiments show that the animal does not have to (and often cannot) taste or smell the presence of the desirable, or undesirable, substance in their food. It relies upon post-ingestive factors to provoke the relevant response.

hunting An aspect of *predatory behaviour that involves a member of one species, the hunter, capturing a member of another species, the prey. A predator may choose a suitable ambush site and sit in wait, it may move about *searching for prey, or it may attempt to flush prey from their place of concealment.

Many carnivores, including lions (*Panthera leo*), spotted hyenas (*Crocuta crocuta*), and jackals (*Canis mesomelas*), scavenge on carrion and steal directly from other predators, in addition to hunting and killing on their own account. Predators, such as the lion and cheetah (*Acinonyx jubatus*), are able to outrun their prey over short distances, and usually precede the chase by a stealthy approach. Such stalking requires inconspicuous manoeuvring, and it often benefits the predator to have *camouflage. Thus, the common cuttlefish (*Sepia officinalis*) adopts a cryptic coloration and approaches its intended victim slowly. Both the pike (*Esox lucius*) and the cheetah make skilful use of cover, moving from one clump of vegetation to another.

There are two main modes of pursuit. During **guided pursuit** the predator continually monitors the position of the prey and alters its pursuit accordingly. Cheetah may sprint to bring themselves with striking distance, and then they follow the zigzags of the prey as they close in. During **ballistic attack** the predator launches itself at the victim, having first estimated its future path. No attempt is made to modify the direction of the attack, once it has been launched. Both large-mouthed bass (*Micropterus salmoides*) and the common octopus (*Octopus vulgaris*) attack in this way. The exact mode of capture is usually tailored to the species of prey. Thus pygmy owls (*Glaucidium passerinum*) kill small birds with the powerful grip of their talons, but kill mice by repeated pecks at the head.

Communal hunting requires a degree of *social organization. Lions hunting as a group kill about twice as many prey per hunt, as do single lions. Communal hunting also enables larger prey to be obtained. Thus killer whales (*Orcinus orca*) individually attack small prey, but they can successfully attack large baleen whales (*Mystacoceti*), when hunting as a cooperative group. Similar cooperation occurs when wolves (*Canis lupus*) attack moose (*Alces alces*), and when cheetah attack zebra (*Equus*). The obvious advantages of communal hunting are partially offset by the fact that the carcass is generally shared among members of the group. Nevertheless, communal hunting increases the predators' range of opportunity. Thus a number of predators, including wolves and killer whales, herd their prey, approaching it with the pack fanned out. Eventually, they encircle the prey, or drive it into a place from which there is no escape. In some species different roles are adopted by different members of the pack. Some drive the prey, while others wait in ambush. In a mated pair of lanner falcons (*Falco biarmicus*), the female chases pigeons (*Columbia livia*) or jackdaws (*Corrus monedula*) away from a sea cliff, by flying in and out of gullies and caves, while the male waits on the wing to attack the birds flushed by his mate.

hypnosis A form of tonic immobility, akin to *death feigning, induced in the laboratory. Animal hypnosis has been known about for a long time. It can be induced by holding the animal, a chicken, lizard, or lamb, down on the ground for a short period. The animal becomes immobile, often with eyes closed. It may then suddenly get up and attempt to escape.

The hypnotic response is almost certainly due to *fear and can be enhanced by the presence of a predator, such as a stuffed hawk. Many predators ignore motionless prey, and death feigning is a common response to predators in the wild.

I

imitation An aspect of *cultural behaviour, which involves the ability to copy aspects of the behaviour of another individual. Apparent imitation may occur as a result of *social facilitation, *emulation, and a tendency to investigate places where other members of the species have been observed. For example, for many years milk was delivered to houses in England, the bottles being left on the doorstep early in the morning. Blue tits (*Parus caeruleus*), and sometimes great tits (*Parus major*), would peck through the foil tops of the bottles and help themselves to the rich cream that floats to the top of the milk. This practice first appeared in particular localities and gradually spread, suggesting that the birds were learning from each other. In fact, a bird does not directly copy another, but learns from observation that the bottle is a source of food. It then learns to open the top by itself.

Direct copying of the behaviour of others occurs in a number of birds and primates. **Vocal imitation** is the copying of members of one's own species, especially when young birds learn their *song from others around them. During their first 90 days of life, young chaffinches

(*Fringilla coelebs*) assimilate the basic structure of their song, although they themselves do not sing. They hear the song of their parents in early life and reproduce it when they are sexually mature, achieving perfection through rehearsal. Many other species of songbird have a similar song development.

Not all cases of vocal imitation develop at an early stage. Mated pairs of the bou-bou shrike (*Laniarius aethiopicus*) have the ability to sing, in rapid alternation and remarkable coordination, the notes of an antiphonal song. Either bird can sing the whole song alone if the other is absent, and both birds can simultaneously and perfectly duplicate each other, as well as sing in alternation. These songs are learned when new pairs are formed and are particular to the mated pair.

Vocal mimicry refers to the apparently non-functional copying that occurs in some species, notably parrots (Psittacidae) and mynah-birds (*Gracula religiosa*). A number of species imitate not only their own species' song, but the songs of other bird species, barking dogs, and human speech. Such mimicry rarely occurs in the wild, but is common in captive birds.

Imitation of movement patterns occurs amongst primates. The celebrated chimpanzee (*Pan troglodytes*) Viki, fostered in a human family, learned to copy human postures and mimic photographs of herself performing particular actions. The question of whether this type of copying merits *mental state attribution is controversial, because of the extensive training involved. Some scientists take the view that true imitation involves *self-awareness. Others dissent from this view.

imprinting An aspect of *learning that takes place during a *sensitive period in the early stages of an animal's life. Lambs (*Ovis ammon aries*) follow the person that has reared them on a bottle. Even after the lamb has been weaned and joined the flock, it will approach its former keeper, and try to stay nearby. If the lamb has grown into a mature male, it may show a sexual interest in its keeper. The lamb is imprinted upon the keeper, and this has both short- and long-term aspects. The lamb follows the keeper when young, and as an adult is shows some attachment to its keeper.

The 'following response' is shown by many precocial species, which can run around soon after birth. The juvenile initially shows a fairly indiscriminate attachment to moving objects. Thus ducklings (Anatidae) separated from their mother will follow a crude model duck, a slowly walking person, or even a cardboard box that is moved slowly away. Some stimuli are more effective than others in eliciting the following response. Thus ducklings prefer to follow yellow-green objects, while domestic chicks (*Gallus domesticus*) prefer blue or orange objects. In general, the more an animal forms an attachment to one object, the less it is interested in others. Attachment can be enhanced by food reward, or the reward of the mother's proximity. One natural function of imprinting is to learn the characteristics of the mother.

Attachment takes place during a sensitive period which varies from species to species. Ducklings most readily form an attachment to a moving object between 10 and 15 days of hatching. If first exposure occurs after this period, the strength of the attachment declines for about another 6 weeks. Imprinting will not occur if first exposure occurs after the sensitive period.

In addition to its effects on parent–offspring relationships, imprinting can have marked effects upon the *social relationships of adults, and on *food selection and *habitat selection. The sexual preferences of many birds are influenced by early experience, a phenomenon called **sexual imprinting**. In ducks (Anatidae), domestic fowl (*Gallus gallus domesticus*), pigeons (*Columbia livia*) and finches (Fringillidae), individuals of one breed with a distinctive coloration can be reared by parents of a different breed and colour. When mature, such individuals prefer to mate with birds of their foster parents' colour, rather than their own colour. Birds that are hand reared often become sexually imprinted upon people, and hand-reared mammals often develop special relationships with people.

In nature, imprinting provides an alternative to *innate recognition of members of one's own species, and of close kin (**filial imprinting**). It also has an important influence upon *mate selection.

incentive An aspect of *motivation that derives from external stimulation. Thus the sight of food (and its apparent palatability) contribute to *appetite or *tendency to eat. The total appetite to eat is made up of *hunger, an internal state, and the incentive that stems from the external stimuli.

incest Sexual intercourse with close relatives. In all human cultures incest is taboo, and for a long time this was thought to be a learned *cultural adaptation, lacking any biological basis. Incest has genetically deleterious consequences and is disruptive of society. Therefore those societies that could ban incest would seem to be better off. Doubts about this type of sociological theorizing began to emerge in the 1970s.

Incest inhibitions occur in a wide variety of animals, including some non-human primates. Moreover, it was observed as long ago as the 19th century that people who grow up with each other experience little mutual sexual attraction, and often a sexual aversion develops. Chinese child marriages, in which the bride is adopted into the grooms family and brought up like a sister (*sim-pua* marriages) result in less sexual interest, fewer children, and a higher divorce rate than adult marriages. Studies of kibbutz children brought up in communal age-classes, show that after puberty such children develop friendly sibling-like relationships, but not sexual relationships. Age-class members do not marry, and it appears that there is a *sensitive period up to the age of six or seven, during which the child learns which individuals are to be sexually excluded, a process called negative *imprinting.

inclusive fitness A measure of *fitness based upon the number of the animal's genes that are present in subsequent generations, rather than the number of offspring. In assessing the inclusive fitness of an animal it is necessary to take account of the number of its genes that are also present in related individuals. This will depend upon the **coefficient of relationship** between the one individual and another. The coefficient of relationship between parent and child is ½ because a child gains half its genes from each parent. The coefficient of relationship between grandparent and child is ¼, and between siblings is ½. *Hamilton's rule defines the mathematical relationship between inclusive fitness and the coefficient of relationship.

incubation An aspect of *parental care in which the eggs are maintained outside the mother's body, in such a way that the embryo can develop and hatch successfully. Such eggs require particular conditions of heat, humidity, oxygen, and carbon dioxide to develop successfully. Some bumble bees (Bombini), some pythons (*Pythonidae*), and most birds sit on the eggs to maintain them in a stable thermal and gaseous environment. Not all birds incubate their eggs, however. Some birds of the mound-builder family (Megapodiidae) build a mound of decaying vegetation in which they lay their eggs. The decomposition of the mound provides heat, and the adult tends the mound until the eggs hatch. Other megapodes lay their eggs where they can be heated by the sun or by underground volcanic activity.

Most birds sit on the eggs during incubation. The optimum temperature for the development of the embryo inside the egg varies from species to species, but is generally within the range 34–39°C. Incubation below the optimum may lead to developmental abnormalities. However, eggs can remain below the temperature at which no development occurs for several days, without ill-effect. The duration of incubation also varies from species to species.

Most birds are attentive to the eggs during incubation, turning them periodically, protecting them from undue sun, and from predators. The parent controls the temperature of the eggs, essentially by including the eggs in their own bodily *thermoregulation. Thus, if eggs are artificially cooled or warmed, the incubating bird will respond accordingly.

individual distance The distance from an individual at which another of the same species provokes *aggression or *avoidance. Some species have virtually zero individual distance, while others are gregarious but still maintain characteristic individual distances. There may be differences between the sexes. Male chaffinches (*Fringilla coelebs*) maintain greater individual distances than do females. Experiments in which female chaffinches are dyed to resemble males show that plumage colour is a major factor. In a group of eight females, disguised as males, the individual distances were the same as males. Thus it seems that if one's neighbour looks like a male, one does not approach too closely.

innate behaviour Behaviour that is inborn in the sense that it has high *heritability. Historically, the concept has been the subject of much debate, and extreme views. One such view was that genetic influences are minimal and that most behaviour develops as a result of learning and imitation. Another extreme view was that behaviour is primarily the result of *instinct, which can be modified by learning in some cases. Many behaviour patterns seemed innate, in the sense that they develop without example or practice. Most scientists now recognize that all behaviour s influenced to some extent by the animal's genetic make-up, and, at the same time, by the environmental conditions that exist during development. The extent to which the two influences, nature and nurture, determine the outcome varies greatly from species to species, and from activity to activity within a species. Examples of these variations can be seen in the study of bird *song.

Much animal behaviour is innate in the sense that it inevitably appears as part of an animal's repertoire under normal and natural circumstances. An activity can be innate in this usage, even though it is learned. For example, the juveniles of many species learn the characteristics of their own species members, and of their *habitat, at a particular stage of development. For this *imprinting to occur in the proper manner, the *parental care must follow its normal pattern. Thus although genetic factors may be necessary for certain behaviour to develop, the species-characteristic processes of *maturation and *learning may be just as important. The naïve dichotomy between nature and nurture, which has characterized much of the debate about innate behaviour, bears no relation to the complex and subtle interactions between the various processes that occur during *ontogeny.

insight learning An aspect of *learning in which supposed insight into the relationships among stimuli or events leads to the sudden production of a new response. For example, a caged chimpanzee (*Pan* sp.) may fit two poles together to reach through the bars of the cage and retrieve a banana placed outside. Similarly, a chimpanzee may drag a box into position and stand on it to reach a banana suspended from the roof of the cage.

Doubt has been cast on the concept of insight, by experiments that show that prior familiarity with sticks and boxes is necessary for a solution to the problem set by the experimenter. Moreover, once an animal discovers that it can achieve a particular manipulation, such as fitting two sticks together, it tends to repeat it over and over again. Once the behaviour becomes established in the animal's repertoire, it can be deployed in a number of contexts. This may be *intelligent behaviour that can be accounted for by ordinary learning processes.

instinct Inborn propensity to behave in a certain way (*see* INNATE BEHAVIOUR).

instrumental conditioning A form of *conditioning in which making a particular response is instrumental in obtaining a reward. It is envisaged that an *association is made between a particular response and a particular *reinforcement situation. Thus, in a laboratory study, a rat (*Rattus norvegicus*) may associate pressing a lever with delivery of food. In nature, a bird may associate turning over a leaf with the discovery of an insect. In both cases the response is instrumental in obtaining food, which is said to reinforce the response. Such reinforcement may be negative. Thus a cow (*Bos primigenius taurus*) that touches an electric fence and receives a shock becomes conditioned not to touch the fence.

In an instrumental conditioning situation the animal is reinforced for performing some act. The reinforcement must occur in some stimulus situation. Particular features of this situation will be associated regularly with reinforcement, thus providing the essential conditions for *classical conditioning. It is now recognized that the behaviour seen in instrumental conditioning situations is, in part, due to classical conditioning.

intelligent behaviour Behaviour having consequences that are judged to be intelligent. The consequences of behaviour are due both to the behaviour itself and the environment that it influences. When these are judged in relation to some criteria of intelligence, both the behaviour and its appropriateness in the environment in which it is observed should be taken into account. Various criteria for judging behaviour in terms of intelligence have been used. Normally, the criteria relate to what we think about human intelligence. In other words, we judge animal behaviour to be intelligent if it is similar to human intelligent behaviour. This is not good scientific practice. For one thing, it is not possible to devise a fair intelligence test for an animal. Not only are some species good at some types of *problem solving, and other species other types, but the performance of animals depends very much upon the testing apparatus. For example, rats (*Rattus norvegicus*) perform much better at problems involving visual discrimination, if they are required to jump at the symbol of their choice (e.g. in a Lashley jumping stand), instead of running at it (e.g. in a Y maze). It is not possible to find the best testing apparatus for a given species, because nor all possibilities can be tried. It is not fair to compare animals tested in different circumstances, nor is it fair to test animals of different species in the same apparatus.

Other criteria for judging the intelligence of the consequences of behaviour include economic criteria (i.e. in terms of the animal's own energy economy), appropriateness or success in nature (i.e. in terms of *fitness), and in terms of theoretical *optimality criteria. In general, in attempting to define intelligent behaviour, what matters is the behavioural outcome, not the nature of the mechanism by which the outcome is achieved. This is particularly true in assessing the role of *cognition in intelligent behaviour. Cognition provides a way of coping with unpredictable aspects of the environment, but we should ask what advantage an animal would gain from a cognitive solution to a problem. A tried and tested rule-of-thumb (a risk-averse solution) may be better than an essentially experimental (risk-prone) cognitive solution. Thus, in judging animal behaviour, we must distinguish between cognition (a possible means to an end) and intelligence (an assessment of performance in terms of some functional criteria).

intentional behaviour Behaviour that has the appearance of being intentional or purposive. **Intention movements** are incomplete behaviour patterns that provide potential information that an animal is about to perform a particular activity. For example, when a bird is about to take off in flight, it first crouches, then raises its tail and withdraws its head. The crouching may occur a number of times before the bird takes off, or it may not precede flight at all. Intention movements may undergo *ritualization and play a role in animal *communication.

Intention movements are not usually regarded as evidence of true intention, but are likely to be the initial stages of behaviour patterns that are terminated prematurely, either because the animal is in a motivational *conflict, or because its *attention is diverted to some other aspect of behaviour. Indeed, it is difficult to imagine how such incipient behavioural fragments could be avoided in an animal with a complex repertoire of activities.

True intention behaviour, that is carried out by reference to an explicit representation of a *goal is difficult to identify in animals. *Deceitful behaviour is sometimes taken as evidence of true intention, but the issue remains controversial.

isolating mechanisms Behavioural and physiological mechanisms that prevent interbreeding and gene exchange between animal species. **Pre-mating mechanisms** are those that prevent interbreeding. These include: (1) ecological (habitat) isolation, in which closely related species live in the same area, but breed in different habitats; (2) temporal isolation, in which related species breed at different times of day or in different seasons; (3) behavioural (sexual) isolation, in which differences in *courtship behaviour and *mate selection prevent interspecific mating; and (4) mechanical isolation, in which morphological differences in the genetalia prevent completion of interspecific mating.

Post-mating mechanisms are those that prevent full success of interspecific crosses. These include: (1) gametic mortality, in which sperm transfer takes place but the egg is not fertilized; (2) hybrid inviability, in which the offspring have reduced viability; and (3) hybrid sterility, in which the offspring are viable but sterile.

juvenile behaviour The behaviour of sexually immature animals. Like adult behaviour, this is subject to *natural selection. Although some aspects of juvenile behaviour are precursors of adult behaviour, other aspects are specifically adapted to the survival of the young animal. For example the *alarm responses of herring gull (*Larus argentatus*) chicks are quite different from those of the parents. When alarmed, the chicks move a short distance from the nest and crouch silent and motionless among the vegetation, whereas the adults fly away, uttering alarm calls.

Typical juvenile behaviour includes *food begging, *play and specialized behaviour such as *hatching. The development of juvenile behaviour is the result of an interplay of *innate behaviour and *learning, especially *imprinting. The emphasis given to these differing processes depends upon the *habitat and lifestyle of the species. Juvenile animals occupy a *niche that is both typical of the species and of the particular stage in the life cycle. Sometimes, the lifestyle of the juvenile is quite different from that of the adult, examples are the tadpoles of frogs (Anura) and the caterpillar larvae of butterflies (Lepidoptera). In other species, such as the wildebeest (*Connochaetes* sp.), the young animal runs with the herd within a few minutes of being born, and has a lifestyle that is almost identical to that of its parents.

j

kin recognition The ability to recognize individuals to which one is genetically related. This is often the result of early experience. For example, Belding's ground squirrels (*Citellus beldingi*) are born underground, but mother and offspring do not learn to recognize each other until the young emerge above ground at 3½ weeks old. Only then will there normally be any chance of confusion. Ground squirrels that are cross-fostered in nature (raised by a foster family following the death of their mother) treat their foster family as their own. This occurs only if the cross-fostering occurs before the squirrels are weaned and emerge above ground. Underlying this *imprinting effect there may also be a genetic effect. If squirrel pups are reared apart and then tested together, siblings are less aggressive than unrelated individuals.

Most ground squirrel litters are fathered by more than one male, and littermates may be full siblings or half siblings. Observations in nature show that female squirrels are more altruistically inclined towards their full siblings than their half siblings. The ability of the female to make this discrimination must be due to some kind of **phenotypic matching**, an ability to compare one's own phenotype with that of others.

kin selection A form of *natural selection that favours genes that promote *altruism towards individuals that are genetically related to the altruist. The degree to which altruistic behaviour should be extended towards other individuals will depend upon the probability of the gene being represented in those other individuals (i.e. upon the *coefficient of relatedness). This does not mean that the individual altruist has to calculate its relatedness to each possible recipient. It may be that one's neighbours are more likely to be one's relatives, as in rabbits (*Oryctolagus cuniculus*). Those rabbits that give the *alarm signal by thumping on the ground both endanger themselves and warn nearby rabbits. These neighbours are more likely than not to be kin, carrying some of the thumper's genes, probably including the one that facilitates thumping. Thus the thumper is benefiting those other individuals that have a tendency to thump on the ground when alarmed.

The alternative strategy is for altruists to recognize their kin and confine their altruistic behaviour to them. Such *kin recognition is known to occur in some species.

kinaesthetic senses The detection of changes within an animal's own body by internal *sense organs.

kinesis A form of *orientation in which the animal's response is proportional to the intensity of stimulation, and is independent of the spatial properties of the stimulus. For example, common woodlice (*Porcellio scaber*) are very active at low levels of humidity and less active at high humidity. Consequently, they spend more time in damp parts of the environment and their rapid locomotion in dry places increases the probability that they will discover less dry conditions. Woodlice tend to form *aggregations in damp places beneath rocks and fallen logs, and their selection of this *habitat is based entirely upon kinesis.

klinokinesis A form of *kinesis in which changes of direction are more common the stronger the stimulus. For example, the flatworm (*Dendrocoelum lacteum*) increases its rate of turning as light intensity increases. Consequently it tends to spend less time in light places, and tends to form *aggregations in dark places.

klinotaxis A form of *orientation achieved by successive comparison of the stimulus intensity in different nearby places. The rate of turning, or sampling the environment, varies with the stimulus intensity. For example, the maggot of the house fly (*Musca domestica*) has primitive *photoreceptors in its head, which enable it to register the light intensity. As the maggot crawls along it moves its head from side to side, testing the light intensity on each side of its body. If the intensity on the right side is greater than that on the left, then the maggot is less likely to turn its head towards the right. It therefore tends to change its course and head away from the source of light. It is thus able to achieve *taxis, a form of orientation in which the animal heads directly towards or away from the source of stimulation.

knowledge The basis of *cognition. Knowledge depends upon *representations. Implicit representations are the basis of know how, or **knowledge how**, sometimes called procedural knowledge. This type of knowledge has to do with accomplishing skilled tasks, such as capturing prey. In everyday life we call this knowing how to do something, such as ride a bicycle.

Explicit knowledge, or **knowledge that**, involves explicit representations of facts that are accessible to a number of processes, and not simply part of a fixed procedure. In human terms, explicit knowledge would imply knowledge of facts that could be put to many uses. Thus knowing that fire is hot can be useful in cooking food, clearing land, repelling predators, etc. Whether animals have explicit knowledge is controversial (*see also* SELF-AWARENESS).

landmarks An aspect of *navigation by which pilotage, steering a course using familiar landmarks, is achieved. The role of landmarks is of importance primarily as the destination is approached. Allowing pigeons (*Columbia livia*) and bees (Apidae) to view familiar landmarks improves their navigation, and disrupting landmarks causes navigational accuracy to deteriorate in many species. Landmarks have been found to be important in the navigation of birds, fish, and many types of insect. It has been suggested that landmarks are identified by means of a 'snapshot'-matching mechanism that enables the animal to identify a string of landmarks when it is proceeding along a familiar route, but this theory is controversial.

language An aspect of *communication, typical of humans, which is characterized by various features, some of which are found in other animals. For example, the signals employed in human language are arbitrary in the sense that they do not physically resemble the features of the world they represent. This abstract quality is also found in animal communication, such as the waggle-dance of honey-bees (*Apis mellifera*). Thus the rate of the waggle-dance is indicative of the distance of food from the hive, the precise relationship being a matter of local convention.

Another feature of human language is that it is an open system into which new messages can be incorporated. The bee dance can be used to direct bees to sources of food, water, propolis (a type of tree sap used to caulk the hive), and to possible new hive sites at swarming time. A number of species give *alarm calls that differ according to the type of danger (*see* Functional reference).

Some aspects of human language, such as the use of grammatical rules, seem to separate it from other aspects of animal behaviour, but even this controversial. Attempts to teach chimpanzees (*Pan* sp.), and other apes, various types of human language have had limited success. Some apes seem to be able to attain a standard similar to that of a young child. The difference may be one of *intelligence, or to the possibility that humans have an *innate language acquisition device.

Thus, although many of the fundamentals of language are not specific to *Homo sapiens*, the question of whether the difference between humans and other animals is one of capacity, or is a genuine structural difference, remains controversial.

latent learning A type of *learning in which the animal appears to make no response to the stimuli that it is learning about. For example, an animal may pass a previously undetected source of food in its way to find water. On a subsequent occasion, when hungry, it will head straight to the food source, showing that it learned about it even though it did not respond to it at the time.

law of heterogeneous summation The independent and heterogeneous features of a stimulus situation are additive in their effects upon behaviour. This law has been found to apply in some cases. For example, the colour markings of the cichlid fish (*Haplochromis burtoni*) are additive in their effect in eliciting aggression in a rival fish. However, the law has not been found to be universally true.

learning An irreversible change in response to particular stimuli, as opposed to the reversible changes that result from changes in *motivation. Other irreversible changes, such as those due to *maturation, injury, and ageing, are regarded as an aspect of *ontogeny. Learning is characterized by changes in *memory of the contiguity of stimuli and of the consequences of responding to stimuli.

The most simple form of learning is *habituation, in which repeated applications of a stimulus result in decreased responsiveness. For example, the escape response of a guppy (*Poecilia reticulata*) to a shadow passed overhead diminishes progressively if the stimulus is presented every 2 min. Eventually the fish does not respond at all.

The type of learning in which an animal forms an *association between a previously significant stimulus and a previously neutral stimulus or response, is called *conditioning. In *classical conditioning an association may be formed between a significant stimulus, such as the sight of food, and a neutral stimulus, such as a flashing light. Initially, the animal responds only to the food (e.g. by salivating). If the food is presented together with the light on a number of occasions, then the animal comes to associate the light with the food, and will eventually salivate even if the light is presented alone. The response of the animal is said to have become conditioned to the light.

In *instrumental or *operant conditioning, the association is made between a particular response and a particular *reinforcement situation. Thus, in a laboratory study, a rat (*Rattus norvegicus*) may associate pressing a lever with delivery of food. In nature, a bird may associate turning over a leaf with the discovery of an insect. In both cases the response is instrumental in obtaining food, which is said to reinforce the response. Such reinforcement may be negative. Thus a cow (*Bos primigenius taurus*) that touches an electric fence and receives a shock becomes conditioned not to touch the fence.

Various other types of learning have been suggested as being involved in *problem solving, notably *insight learning. This involves supposed insight into the relationships among stimuli or events that leads to the sudden production of a new response. For example, a caged chimpanzee (*Pan* sp.) may fit two poles together to reach through the bars of the cage and retrieve a banana placed outside. Similarly, a chimpanzee may drag a box into position and stand on it to reach a banana suspended from the roof of the cage. However, doubt has been cast on the concept of insight, by experiments that show that prior familiarity with sticks and boxes is necessary for a solution to such problems. Moreover, once an animal discovers that it can achieve a particular manipulation, such as fitting two sticks together, it tends to repeat it over and over again. Once the behaviour becomes established in the animal's repertoire, it can be deployed in a number of contexts. This may be *intelligent behaviour that can be accounted for by ordinary learning processes.

Learning occurs when a novel stimulus is accompanied by an event of consequence to the animal, or when the animal's own behaviour is followed by some event of consequence. When the relationship between stimulus and behaviour is no longer followed by relevant consequences, then *extinction occurs and the animal gradually ceases to make the learned response. For example, if an animal discovers food in a particular place, it may learn to visit that place on future occasions, provided it finds food there, at least some of the time. If the animal no longer finds food, it will visit the place less and less frequently, until the learned behaviour is extinguished, and the behaviour ceases entirely. Extinction is not simply a matter of lack of *memory. Relearning after extinction is usually much more rapid than the original learning, which suggests that the process of extinction does not abolish the original learning, but somehow suppresses it. Further evidence for this conclusion comes from the phenomenon of spontaneous recovery, by which a response that has been extinguished recovers its strength with rest, during which no relevant stimuli are present.

Learning by association implies that there is a linkage between stimuli, rather than a linkage between stimulus and response. Different theories of association may ascribe the linkage to stimulus substitution (one stimulus standing for another); contiguity of stimuli (contiguous stimuli are associated); or the learning of cause and effect relationships, in which an event (the cause) can cause another event (the effect) to happen, or not to happen. These different theories of association form part of the subject usually known as animal learning theory.

The strength of an association between a stimulus and some consequences depends partly upon the nature of the consequences. This principle is known as *stimulus relevance. Thus a rat (*Rattus norvegicus*) readily associates a light or sound with electric shock consequences, but does not associate taste stimuli with such consequences. On the other hand, the rat strongly associates taste stimuli with subsequent poisoning, but does not associate light or sound stimuli with such consequences.

When an animal has learned a particular response to a particular stimulus, it may also show the response to other similar stimuli. This phenomenon, known as stimulus *generalization, was first described by *Pavlov. The normal explanation of stimulus generalization is called the common elements theory. A stimulus used in a *conditioning experiment consists of a collection of separate elements. For example, a tone has a given frequency, intensity, and duration. These dimensions of the stimulus may become conditioned during acquisition of a conditional response. A new stimulus that has some elements in common with the conditional stimulus may be capable of eliciting the conditional response to some extent. Whereas humans readily distinguish between a 1000 Hz tone and a 300 Hz tone, pigeons (*Columbia livia*) treat them as similar. The tones have other qualities in common, especially the fact that they are unlike natural sounds, in being characterized by a single frequency. It is perhaps not surprising, therefore, that pigeons treat them as similar. If a pigeon is rewarded in the presence of a 300 Hz tone, then it has a lesser tendency to respond to tones of other frequencies.

leisure activities are those that disappear from an animal's repertoire when demands upon the animal's time are very severe. An animal normally spends a certain amount of time each day feeding, sleeping, defending its territory, etc. The amount of time it needs to complete each task depends upon the prevailing environmental circumstances, and upon the animal's state of *motivation. For example, when food is scarce, an animal will have to spend more time *foraging than when food is plentiful.

When *demand upon an animal's time is severe, certain aspects of its behaviour will be curtailed. These will generally be the less essential activities, such as sleep, grooming, and play. When time is very short, some of these less important activities may be abandoned altogether. These are the leisure activities. In other words, leisure activities are those that show the least *resilience to the pressures of time.

In artificial environments, such as zoos, where animals are kept in captivity, an animal may have too little to occupy its time. Such animals normally have their food provided, and some activities, such as *territorial behaviour may be totally precluded. When an animal's daily activities take up less time than is natural, it will generally *sleep more, and may develop symptoms of *boredom.

lek A type of *territory held by males of certain species, and used solely as a communal mating ground. For example, the sage grouse (*Centrocercus urophasianus*) inhabits the high sage bush plains of North

America. The female has a brownish cryptic coloration, while the male, when in breeding condition, is conspicuous. He is larger than the female, and has yellow air sacks, a white breast, and spiky tail feathers. Large numbers of males aggregate in the breeding season and take up breeding territories, or leks, in which there is a dominant male and several subordinate males. It appears that the females are choosing a suitable mate when they move around and between leks, prior to copulation.

The majority of copulations are performed by dominant males, and it appears that females choose mates on the basis of their ability to hold leks. (The males and females live apart most of the year and only the females care for the young.) Some copulations are performed by subordinate males, when the dominant male is distracted by rivals.

Leks are found in many species of grouse (Tetraoninae), and also among pheasants (Phasianinae), hummingbirds (Trochilidae), some bowerbirds (Ptilonorhynchinae), some manakins (Pipridae), and some weavers, such as the village weaver bird (*Textor cucullatus*). Many mammals have communal mating grounds, but the lek type, with an element of female choice, is rare, although it has been observed in the Ugandan kob (*Adenota kob thomasi*).

locomotion Movements that displace the whole body of the animal. Some animals are restricted to one type of locomotion, while others have alternative possibilities, such as *walking and *flight.

In water, many small organisms are free floating and are carried hither and thither as the water moves. A great many of the small animals of the plankton, even those that have means of locomotion, are moved more by water movements than by their own efforts. Modes of active movement in water include swimming, walking (as in crabs (Brachyura)), rowing, hydrofoil swimming, and jet propulsion.

In rowing on the surface of the water, the oar is moved backwards through the water, propelling the body, and then lifted out of the water for the return stroke. In underwater rowing, the oar blades are held broadside during the backward stroke and sideways, or collapsed, during the forward stroke, as in swans (*Cygnus* sp.). In water beetles (*Dytiscidae*) the third pair of legs are the principal oars. They are fringed with long hairs, which are hinged and spread out during the back stroke and trail behind during the return stroke.

Hydrofoil swimming is found in tunnies (*Thunnus* spp.) and other fast-swimming fish. The fish are propelled by hydrofoil tails that beat from side to side. In every stroke the angle of attack of the hydrofoil is adjusted in such a way that the resultant force has a forward component. The left and right components cancel each other out, leaving only the forward components. Hydrofoil swimming is also found in whales (Cetacea), turtles (Chelloniidae), and penguins (Spheniscidae).

Jet propulsion is found in squids (*Loligo* sp.) and octopods (*Cephalopoda*). They draw water through a wide opening into a cavity that has muscular walls, and then squirt it out again through a much smaller opening, the funnel. They can point the funnel forwards or backwards. Dragonfly larvae (Anisoptera) also use jet propulsion in their escape movements. They squirt water from the anus to propel themselves forward.

On land, locomotion is achieved by various types of crawling and walking, and other types of legged ambulation. In addition, there are specialized types of locomotion underground and in trees. Locomotion by gastropod molluscs (Gastropoda) and segmented worms (*Annelida*) is achieved by **concertina crawling** over surfaces both above and below water. That part of the body that rests on the ground alternately lengthens and shortens. In its simple form, as in leeches (*Hirudinea*), the whole body lengthens and shortens. The leech has a sucker at each end of the body. When the posterior sucker is attached, the body lengthens, and, while the anterior sucker is attached, it shortens, as so the animal progresses forwards. The same principle of lengthening in front of an anchored region and shortening behind one is used by many worms. Underground, earthworms (*Lumbricus* sp.) burrow using the same technique as they use to crawl. Each segment advances while it is thin, so the anterior end can be insinuated into narrow crevices. The thick segments are firmly jammed in the burrow, while the thin ones advance. Snails (Gastropoda) and other gastropod molluscs have a large, flat foot which rests on the ground. Waves of muscle contraction travel along it, throwing the sole into ripples, providing the same principle of propulsion as that found in worms.

Snakes (Serpentes) use various techniques of crawling, of which **serpentine crawling** is the most usual. Muscular waves travel backwards along the body. Stones, grass tussocks, and other irregularities in the ground prevent sections of the body from sliding broadside, but allow them to slide lengthwise, and so the body moves forward. Some snakes (*Serpentes*) climb through trees using the same technique.

Lorenz, Konrad Zacharias (1903–89) Austrian animal psychologist who, together with Niko *Tinbergen, founded modern ethology. He studied medicine in Vienna and became a demonstrator and then lecturer in comparative anatomy and animal psychology. In 1940 he was appointed Professor of Philosophy at the University of Konigsburg, but in 1943 he was drafted into the army medical service. In 1944 he was taken prisoner by the Russians. He was released in 1948 and became attached to The University of Munster. He then moved to Seewiesen with the founding of the Max Planck Institute for Behavioural Physiology, where he remained until his retirement in 1973. That year, together with Niko Tinbergen and Karl von *Frisch, he was awarded the Nobel Prize for Medicine.

lunar rhythms *Rhythmic behaviour that is entrained to the movement of the moon and the tides. The lunar cycle of 29.5 days is known to influence a variety of aspects of behaviour, primarily in marine animals. For example, the palolo worm (*Eunice* sp.) releases its gametes into the sea only during the neap tides of the last quarter of the moon in October and November. Worms isolated in the laboratory produce their gametes at the correct time, showing that this activity is probably controlled by an endogenous *biological clock. Amongst crabs (Brachyura) of various species, the tidal rhythm is superimposed upon a *circadian rhythm, resulting in nocturnal activity only when the tide is high.

M

magnetic orientation An aspect of *orientation that depends upon detection of elements of the Earth's magnetic field. Responses to magnetic fields have been shown to occur in many species, both in the laboratory and in the field. Miniature magnets have been found in bacteria, honey-bees (*Apis Mellifera*), and pigeons (*Columbia livia*), but how the magnetic information is made known to the nervous system remains to be discovered. In birds, it appears that it is the declination of the Earth's magnetic field that is important in providing a directional sense, and it would seem that this is used in combination with visual cues as an aid in *navigation.

male rivalry An aspect of *sexual selection in which males compete, directly or indirectly, for access to females. Amongst red deer (*Cervus elaphus*), for example, the females have little or no choice of sexual partner, because the males defend their harems against possible rivals. Competition between two stags involves *assessment and sometimes *fighting.

Some contests between males involve mock fighting. Thus buffalos (*Syncerus*) and bighorn sheep (*Ovis canadensis*) charge each other and clash head on in *ritualized trials of strength. In other species the fighting is more serious. Among elephant seals (*Mirounga angusirostris*) male rivalry is intense and often involves fighting. When a male attempts to copulate with a female, she protests loudly, attracting the attention of other nearby males, who attempt to interfere. A male is likely to be successful in copulating only if he is dominant and can ward off his rivals. The female intensifies the competition by her protests and ensures herself a dominant male. Thus while the female does not directly exercise a choice, her behaviour indirectly has that effect. If the female does not protest, the copulation is less likely to be interrupted and a low-ranking male will have a greater chance of success.

manipulation An aspect of *communication, which can occur when the receiver obtains information about the signaller, against the interests of the signaller. For example, the courtship display of a male may attract rivals. The rival may benefit by receiving the signaller's message to the female, because it leads him to the female. This form of manipulation has been called **eavesdropping**.

mate selection The initial stage of the reproductive cycle, in which a member of the opposite sex is singled out for attention. Mate selection is often followed by *courtship. Normally, mate selection occurs among sexually mature individuals, but there are some exceptions to the rule. Among some lovebirds (*Agapornis*) heterosexual pairs form when the birds are about two months old and still have their juvenile plumage. In the African waxbill (*Uraeginthus granatinus*) monogamous pairs are established before the partners are 35 days old, and are still being fed by their parents.

An important aspect of mate selection is selecting the right species and sex. When parents are of different species, the offspring are generally infertile and incapable of adapting to the ecological *niche of either parent. Hence hybridization in nature is invariably disadvantageous, and the pressures of natural selection generate mechanisms designed to avoid such misidentification. The barriers to hybridization, whether geographic, climatic, mechanical, or behavioural, are known as reproductive *isolating mechanisms.

Mates must be selected according to sex as well as species. While in many species males and females differ in a number of ways, sexual identification sometimes depends upon the identification of relatively few features. For example, in many species of pigeon (Columbidae) the main distinguishing feature is behavioural. The male performs a bowing display when he encounters a member of the same species. The sexes differ in their response to this behaviour.

In many promiscuous species, it is important to select a mate that is physiologically ready for mating, otherwise much time and effort can be wasted. Females often indicate their readiness for *sexual behaviour by means of pheromones, but this does not necessarily mean that they will cooperate. Females are often 'coy', testing the male by prolonging the courtship. Even when the female permits copulation, it does not necessarily follow that the male will father her offspring. In many species the female harbours the sperm from a number of males, and is able to exercise some degree of control of the fertilization process. Thus *sperm competition is the final arbiter of mate selection.

maternal behaviour An aspect of *parental care, confined (by convention) to female mammals that suckle their young from specially developed mammary glands. Maternal behaviour is that behaviour, exhibited by mothers towards their young, which is presumed to aid the young in their survival, growth, and development, both physically and behaviourally. The specific maternal behaviour exhibited by the different species of mammal is quite varied. The three main types of maternal behaviour are: (1) that shown by mothers of **altricial** young that are helpless at birth; examples are found among dogs (Caninae) and rodents; (2) that shown by mothers of **precocial** young that can see and hear, thermoregulate, and walk soon after birth; examples are found among sheep (*Ovis*), deer (Cervidae), and cattle (Bovidae); and (3) that shown by mothers of semi-helpless young that cling to their mother's fur; examples are found amongst the primates.

mating systems A fundamental aspect of *social organization. The two basic types of mating system are *monogamy and *polygamy. In monogamous systems each breeding adult mates with only one member of the opposite sex. Monogamy is common in birds but rare in other animals. In **perennial monogamy** a individual mates with one partner for life, or until a 'divorce' takes place as a result of failure of one partner to fulfil its duties. The partners usually maintain some association even outside the breeding season. Such relationships are found amongst swans (*Cygnus* sp.), field geese (*Anser* sp.), cranes (*Grus* sp.), and gibbons (Hylobatidae).

Seasonal monogamy is characteristic of species that are monogamous during the breeding season, but lead separate lives for the rest of the year. Many migrant birds fall into this category, and they usually have a strong tendency to return to their exact nesting locality, where they join up with their partner each season. Herring gulls (*Larus argentatus*) are an example.

In polygamous species an individual generally has two or more mates, either successively or simultaneously. The most common form is **polygamy**, in which one male mates with a number of females. In **polyandry** one female mates with a number of males. An example is provided by the American jacana (*Jacana spinosa*). Here the female is more conspicuous, dominant, and territorial than the male. The female lays the eggs, which are incubated solely by the male, while the female attempts to persuade more males to incubate successive clutches.

In **promiscuous** species there are no pair bonds and both males and females mate with more than one member of the opposite sex. The formation of pair bonds, and the type of mating system that is characteristic of a species, has developed by *evolution in relation to the animal's ecological *niche, and in response to the genetic interests of the sexes. *mate selection, *parental investment, and *territory are all factors that influence the evolution of mating systems.

maturation An irreversible part of *ontogeny. Maturation does not depend upon experience, and is thus distinct from *learning and injury. For example, pigeons (*Columbia livia*) have been reared in confined conditions so that they cannot move their wings. When released, these birds could fly just as well as other birds of the same age. Although young birds can be seen flapping their wings while remaining on the ground, this apparent *practice has no effect upon the development of flight. Similarly, tadpoles (Anura) flex their tails in a coordinated manner while still within the jelly-like spawn. This behaviour can be inhibited by application of a light anaesthetic. If, when the tadpoles have grown to normal swimming size, the anaesthetic is removed, the tadpoles swim normally, showing that the development of swimming movements is one of maturation.

mechanoreceptors Receptors that are sensitive to mechanical deformation. They provide the basic elements of a variety of senses, depending upon where in the body they are situated, and which accessory structures are associated with them. The *orientation of animals in relation to gravity and external stimuli depends partly upon information about the spatial relationship of the various parts of the body. In mammals this information comes from the vestibular system of the ear, and from receptors in the joints, muscles, and tendons. The joint receptors provide information about the angular position of each joint. In the tendons are the Golgi tendon organ receptors, which are sensitive to tension, and within the muscles are the muscle spindles, which are sensitive to changes in muscle length.

Mechanoreceptors deep inside the body serve a wide variety of functions, including detection of changes in blood pressure, bladder pressure, etc. Pacinian corpuscles respond to deep pressure, while Meissner's corpuscle and the nerve plexus around hairs are responsible for touch.

All animals have mechanoreceptors of some kind. Thus crabs (Brachyura) have receptors that signal both position and movement at a joint, and blowflies (*Phormia regina*) have stretch receptors in the gut that inhibit feeding when the gut is distended.

memory Processes that cause information, or behaviour, to persist over time. Memory processes usually occur in the brain, but some types of memory may exist outside the body (*see* STIGMERGY). Memory usually refers to the persistence of learned behaviour over time, in contrast to the persistence of *innate behaviour.

Studies of memory in animals have three distinct aspects: (1) the formation of a habit or response through *learning; (2) the retention of the habit in memory; and (3) the expression of the response in some type of retention test. The basic idea is to study the retention process. During the retention period the experimenter usually introduces certain variables in an attempt to investigate the mechanisms involved in the retention process. For example, if during the retention period for stimulus A, a new stimulus B is introduced, the memory of A (as measured by a retention test) may be inhibited or reduced.

The two basic theories of memory are: (1) that memories decay over time, unless some type of rehearsal takes place; and (2) memories do not decay, but are interfered with by subsequent memories. In other words, there is competition between memories. Thus **forgetting** may be seen either as natural decay, or as the result of interference.

Monkeys (Simiae) and pigeons (*Columbia livia*) can be trained to match samples. A visual stimulus is presented alone, then after a retention period with no stimulus, an array of stimuli is presented, and the animal is rewarded for choosing the one that it saw initially. Memory of the initial stimulus is better if the animal experiences darkness during the retention interval, suggesting that lack of distraction aids memory. Pigeons can remember the sample stimulus for about 10–15 s, while some monkeys can remember the sample for 8–9 min. In addition to such short-term memory, many animals exhibit a longer-term **state-dependent memory**. They remember stimuli relevant to a particular *motivational state, such as hunger or fear, when they are next in that state some hours later, but seem unable to retrieve memories when they are not in the relevant state.

Long-term memory varies with species and with the type of behaviour that is relevant. Thus, rats (*Rattus norvegicus*) remember responses to stimuli relevant to *avoidance and to *orientation, even when tested months later. They do not have good long-term memory for arbitrary responses that they learned to obtain food.

menotaxis A form of *telotaxis that involves *orientation at an angle to the direction of stimulation. Examples are found in the time-compensated sun-compass *navigation of many insects, fish, and birds.

menstrual cycle The repeated pattern of periodic bleeding from the uterus and vagina, that occurs in sexually mature women, in all species of apes, and in some monkey (Simiae) species. The bodily changes occur in response to fluctuations in the levels of circulating *hormones. Animals with a menstrual cycle are usually sexually receptive most of the time, in contrast to those with an *oestrous cycle that are receptive only at the time of ovulation.

mental state attribution The attribution of mental states, such as beliefs, to other individuals. The theory that this occurs in animals, particularly in studies of *deceit, is a controversial one. The evidence is equivocal.

migration Movement of animals from one *habitat to another. Migration between two different habitats on a periodic or seasonal basis is shown by many species, including butterflies (Lepidoptera), locusts, salmon, birds, bats (Chiroptera), and antelopes (Bovidae). In some species, periodic migrations occur in response to changes in environmental conditions. For example, the desert locust (*Schistocerca gregaria*) inhabits seasonally arid areas. Depending upon their degree of crowding, these insects may develop into one of three types of adult. Of these, the *gregaria* individuals aggregate into dense swarms and migrate downwind into areas of low barometric pressure, where they are most likely to encounter rain. They fly by day and stop migrating when they encounter wet conditions. Sexual maturation, copulation, and deposition of eggs then occur. Locust swarms describe seasonal circuits, but the generation time is too short for individuals to complete the entire circuit.

Some seasonal migrants initiate their migration on the basis of a *circannual rhythm, rather than responding to environmental change. European populations of migratory warblers (*Sylvia* spp.) winter in Africa and migrate north across the Sahara desert. When they are maintained under constant laboratory conditions in Europe, having been hand-reared from a few days after hatching, they show periodic migratory restlessness, at the same times that migration is occurring in their wild relatives.

In addition to seasonal migrants, some species undertake migrations as part of exploration and colonization of new habitats. For example, the European collared dove (*Streptopelia decaocto*) has expanded its range westwards over decades. Prior to 1930 the species was primarily an Asiatic bird, not extending further into Europe than the Balkans. Since then it has colonized the whole of central Europe and southern Scandinavia, reaching Britain as a breeding bird in 1955.

The migratory behaviour of animals over vast distances involves considerable feats of *navigation. Some animals, such as herring gulls (*Larus argentatus*) and salmon (*Salmo* sp.), are able to relocate their precise territories with considerable accuracy.

mimicry (*see* IMITATION for another meaning) The resemblance of one animal (the mimic) to another animal (the model) such that the two are confused.

In **Batesian mimicry** the predator avoids a noxious animal producing a particular signal (the model), and is deceived into avoiding an edible mimic which produces a similar signal. It is of advantage to the mimic, but no advantage to the model. The model is usually an animal with a nasty taste, a sting, or spines, but plants and inanimate objects can also be mimicked. For example, stick insects (Phasmida) closely resemble twigs that are inedible to predators.

Batesian mimicry is of advantage to the mimic, but is of no advantage to the model. A limitation of this type of mimicry is that the mimic must not become too common relative to the model. Otherwise predators will sample mimics first, and then develop a conditioned *avoidance to both mimic and model. Mimics gain greatest protection from predation when they are rare in relation to models.

In **Müllerian mimicry** a group of noxious species share the same *warning signals. Examples are the black and yellow social wasps (*Vespula* sp.), the solitary wasps of the genera *Bembex*, *Eumenes*, and *Crabro*, and the cinnabar moth caterpillar (*Callimorpha jaçobaea*). Birds that have learned to avoid cinnabar caterpillar will not attack wasps, but birds with no experience of wasps or cinnabar caterpillars will attack wasps. Since predators need to learn one pattern in order to avoid all species in the assemblage, Müllerian mimicry is of benefit to all individuals and no deception is involved.

In **aggressive mimicry**, one animal (the mimic) preys upon, or otherwise exploits, another organism because it is attractive to, or not avoided by, its prey. For example, the Malayan praying mantis (*Hymenopus bicornis*) is pink and rests among flowers of *Melastoma polyanthum* and closely resembles them in colour and shape. Insects attracted to the flowers are caught by the mantid.

In **intraspecific mimicry** the model and the mimic, and the animal perceiving the mimicked signal, all belong to the same species. For example, the female of the African cichlid fish (*Haplochromis burtoni*) broods her eggs in her mouth until they hatch. This protects them from being eaten by predators. When ready to copulate, the male digs a small scrape in the substratum and attracts the female to this place. She spawns here and immediately takes all the eggs into her mouth. The male then ejects his semen into the scrape. At the same time he expands his anal fin, which has orange circles on it of the same size and colour as the eggs. The female attempts to suck these dummy eggs into her mouth, and in so doing takes in the semen, which fertilizes the eggs in her mouth. This mimicry is of benefit to both male and female, since it increases the chances of successful fertilization (*see* Figure 6).

mixed strategy An *evolutionarily stable strategy in which equilibrium is achieved when each player successively mixes the

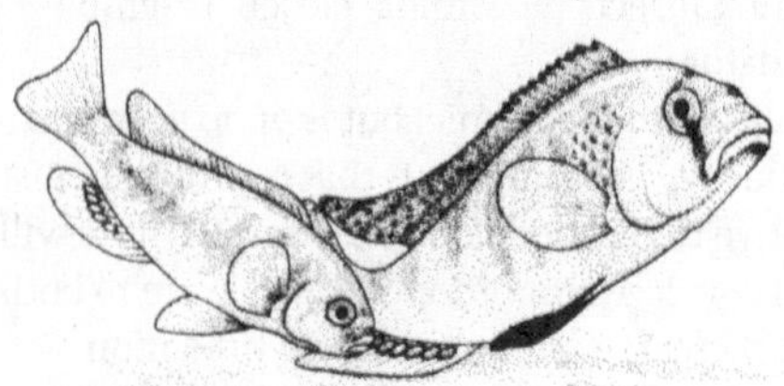

Fig. 6. Male *Haplochromis burtoni* (right) ejecting semen while the female (left) attempts to pick up dummy eggs on his anal fin, and in so doing takes up the semen.

alternative strategies. For example, in the hawk-dove game, the two contestants start in identical situations, and have the same choice of strategies and the same prospective payoffs. If the proportion of hawks in the population is h and the proportion of doves $(1-h)$, the average payoff for a dove will depend upon the probability of meeting a hawk or another dove. However, the same payoff would be achieved by meeting an *individual* that played hawk h per cent of the time and dove on all other occasions. Such an individual would be playing a mixed strategy. It is the ratio of strategies played that matters in the establishment of an ESS, and it makes no difference whether this is achieved by a proportion of a population always adopting one strategy (and the rest the other strategy), or all members of the population playing the equivalent mixed strategy.

mobbing A form of harassment directed at predators by potential prey. For example, owls (Strigiformes) that appear in the daytime are often mobbed by small birds. The birds make mock attacks upon the owl, alternatively approaching and retreating, and uttering a characteristic *vocalization. The mobbing calls of birds are characteristically easy to locate, and serve to recruit would-be attackers.

Mobbing is also found amongst mammals, particularly the primates. Baboons (*Papio* sp.) and chimpanzees (*Pan* sp.) launch mock attacks on leopards (*Panthera* sp.), screaming, charging, and retreating. Californian ground squirrels (*Citellus beecheyi*) mob snakes (Serpentes), as do agoutis (*Dasyprocta* sp.).

The survival value of mobbing lies partly in the *warning given about the presence of a predator, and it is sometimes successful in driving the predator away. A predator that is mobbed has lost the advantage of surprise attack, and would be better off hunting elsewhere.

monogamy A type of *mating system in which each adult mates with only one member of the opposite sex. In **perennial monogamy** a individual mates with one partner for life, or until a 'divorce' takes place as a result of failure of one partner to fulfil its duties. The partners usually maintain some association even outside the breeding season. Such relationships are found amongst swans (*Cygnus* sp.), field geese (*Anser* sp.), cranes (*Grus* sp.), and gibbons (Hylobatidae).

Seasonal monogamy is characteristic of species that are monogamous during the breeding season, but lead separate lives for the rest of the year. Many migrant birds fall into this category, and they usually have a strong tendency to return to their exact nesting locality, where they join up with their partner each season. Herring gulls (*Larus argentatus*) are an example.

Monogamy is usually found in those species in which both parents are required to raise the young. It is common in birds, and occurs in some primates.

motivation A reversible aspect of the animal's state that plays a causal role in behaviour. Changes in behaviour in an unchanging environment may be due to irreversible processes such as *learning, *maturation, or injury, or reversible motivational processes.

An animal's motivational state changes continually as a result of both external and internal changes. External changes may be exogenous, emanating from changes in the outside world that are perceived by the animal, or endogenous and stemming from the animal's own behaviour. In both cases the motivational state of the animal is changed as a result of stimuli impinging on the animal's *sensory systems. For example, potential food items may present themselves to an animal, or they may be revealed as a result of the animal's own behaviour. Thus, an insect may happen to cross the path of a foraging bird, or the bird may discover an insect by turning over a leaf.

Internal changes result both from environmental forces and from the animal's own behaviour. External temperature changes, for example, may affect the animal's motivational state in a purely physical manner, or they may be detected by the animal and cause a change in behaviour that itself causes changes in motivational state. For example, a sudden rise in temperature may cause an animal to pant. The consequent evaporative cooling alters the motivational state, as does the loss of water caused by panting. All behaviour has some effect upon motivational state, even if only because all behaviour results in energy expenditure.

Animals may be divided into **conformers**, which allow their internal environment to be influenced by external factors, and **regulators**, which maintain their internal environment in a state that is largely independent of external conditions. The process by which regulators control their internal state is called *homeostasis. In the case of *thermoregulation, birds and mammals are regulators and most other animals are conformers.

The **motivational state** of an animal is made up of all those internal and external factors that have a causal effect upon behaviour. These include factors relevant to incipient activities, as well as the animal's current behaviour. The combination of external stimuli that give rise to a particular activity is generally known as the *cue strength for that behaviour. For example, the colour markings of the cichlid fish (*Haplochromis burtoni*) are additive in their effect in eliciting aggression in a rival fish. In other words, there is *heterogeneous summation. The stimuli combine to indicate the potential threat. The magnitude of this combination is the cue strength for threat. The body posture of the fish is another matter, since it signifies its aggressive intention. A head-down posture is usually adopted just prior to attack. A multiplicative relationship exists between the stimuli indicating the aggressive potential of a rival and those indicating his aggressive intention. In addition, there may be internal factors, such as hormone levels, which go to make up the total motivational state that is relevant to aggression. There will also be external and internal stimuli relevant to other activities, such as fleeing. The fish cannot both attack and flee simultaneously. Therefore some *decision has to be made between the tendency to attack and the tendency to flee.

It is a common assumption that internal and external stimuli combine multiplicatively so that the tendency to perform as activity would be zero if either the internal cue strength or the external cue strength were zero. While this appears to be true in some cases, this question remains an entirely empirical one, and there can be no general rule.

The consequences of an animal's behaviour influence its motivational state in a number of different ways. Thus two important consequences of *foraging are the expenditure of energy and the intake of food. Energy and other physiological commodities, such as water, are expended in all behaviour to a degree that depends upon the level of activity, the weather, etc. Food intake may alter the animal's state in a number of ways. The presence of food in the mouth may increase appetite temporarily; it may have a satiating effect, or it may lead to the rejection of the food due to its unpalatability. The presence of food in the gut may have a short-term satiating effect, and it may influence other physiological factors, such as temperature and water balance. The food in the gut is digested and absorbed into the bloodstream. The consequent changes in the blood are complex and may influence various aspects of the animal's motivational state. Feeding is not special; all aspects of behaviour have consequences of equivalent complexity. These may include effects upon other animals, and upon aspects of the external environment, such as a nest.

The motivational state of an animal can be described mathematically, in terms of a state space. The consequences of behaviour are continually changing the state, and describe a trajectory through the space.

motivational disinhibition The removal of inhibition in a motivational context. Normally, an animal has *motivation for more than one activity at a time. The dominant motivational tendency is the one that controls the ongoing behaviour. The dominant tendency is said to inhibit the tendencies for other aspects of behaviour. Removal of this inhibition is said to disinhibit the ensuing behaviour. This type of disinhibition is normally defined operationally, thus—the time of occurrence of a disinhibited activity is independent of the level of causal factors relevant to that activity.

motivational state *See* Motivation.

muscles Effector organs made up of complex protein molecules that are capable of contraction and relaxation. Nerve endings at the **neuromuscular junction** set up electrical potentials that cause the muscle to contract. Muscular relaxation results from lack of stimulation. When the muscle contracts it becomes shorter, provided it is not prevented from doing so by being fixed at each end. A muscle can lengthen when it is relaxed, but only if it is stretched by the action of other muscles, or some extraneous force. Muscles are usually arranged in opposing groups that act against each other.

N

natural selection The process by which *evolution of living organisms occurs. Evidence for evolution comes from the fossil record, comparison of present-day species, the geographical distribution of species, and some observations of evolution in action.

The theory that evolution results from the process of natural selection is due to Charles *Darwin. The elements of Darwin's theory can be stated as follows: Within any population of organisms of the same species, there is considerable variation among individuals. Much of this variation has high *heritability. Many more individuals are born (in the case of animals) in each generation than survive to maturity. Therefore the likelihood of an individual surviving to maturity will be affected by its particular traits, especially those that it has inherited from its parents. If the individual survives to maturity and reproduces successfully, its offspring will tend to perpetuate the inherited traits within the population.

Thus the *survival value of a trait is determined by natural selection. That is, the extent to which a trait is passed from one generation to the next, in a wild population, is determined by the breeding success of the parent generation, and the value of the trait in enabling the individuals to survive natural hazards, such as food shortage, predators, and sexual rivals. Such environmental pressures can be looked upon as selecting those inheritable variations that best fit or adapt the animal to its environment. This is what Darwin meant by 'survival of the fittest'.

navigation A form of *orientation in which the animal covers large distances, moving from one location to another, in a goal-oriented manner. Such long-distance orientation has three aspects: (1) **pilotage**, steering a course using familiar landmarks; (2) **compass orientation**, the ability to head in a given compass direction without reference to landmarks; and (3) **true navigation**, the capacity to orient toward a *goal, such as a home or breeding ground, regardless of its direction, by means other than recognition of landmarks. The long-distance *migration of fish and birds usually involves all three types of orientation.

Two main types of mechanism can be employed in navigation. In **bicoordinate navigation** two coordinates (e.g. latitude and longitude) are used (indirectly) to form a grid. Thus, if we know the latitude and longitude of our present position and also the latitude and longitude of the goal, we can plot a course from the one to the other. In theory, information gained by observing the movement of the sun, or the pattern of stars on a cloudless night, is sufficient for bicoordinate navigation, but there is no conclusive evidence of these abilities in animals. There is evidence that navigating birds use the sun and stars as a compass, but a compass alone is not sufficient for bicoordinate navigation. Similarly, there is evidence for use of a magnetic compass by animals, but this alone is not sufficient for bicoordinate (i.e. map and compass) navigation. Despite much research, there is no convincing evidence for the use of a map analogue by animals.

Navigation by **non-bicoordinate** methods is possible in theory, especially in the case of return navigation to a location previously experienced. Information gained on the outward journey can be used to navigate during the return journey. For example, *Cataglyphis* ants of the Sahara desert navigate by path integration. While foraging away from home, they continually measure all distances covered and all angles turned, and integrate these linear and angular components into a continually updated vector that points towards home. *Cataglyphis* uses a compass to monitor the angular components of its movements. This compass utilizes the pattern of polarized light in the sky. The photoreceptors in a particular region of the ant's (and some bee's (Apidae)) eye are sensitive to directional oscillation of this pattern. The pattern of polarization changes with the elevation of the sun above the horizon. The ant is equipped with a neural template that resembles the polarization pattern when the sun is at the horizon, but differs from it for all other elevations. The best possible match between the template and the external pattern is achieved when the insect is aligned with the solar (or anti-solar) meridian. As the animal rotates about its vertical body axis, the match decreases systematically. The best match gives the zero point of the compass, and this deteriorates as the ant selects other compass directions.

Because there is a discrepancy between the internal template and the external pattern, mismatches occur whenever only parts of the external pattern are available. However, as the ant forages for short periods only, it does not matter that the basic setting of the compass is not always accurate, because it is changes in relation to the temporary (if arbitrary) reference point that are important. The foraging ant simply has to integrate, within one foraging trip, the angular deviations that it makes and the associated distance travelled since it left home. Thus the animal has an automatic device that points towards home throughout its wanderings.

It is possible that similar systems exist in other navigating animals, such as homing pigeons (*Columbia livia*). There is some evidence that pigeons make use of changes in the Earth's magnetic field, and in olfactory information, gained during the outward journey. Olfaction is also known to play an important part in navigation by migrating salmon (*Oncorhynchus* sp.) and newts (*Taricha* sp.).

needs Consequences of behaviour that are essential for survival. Thus, an animal needs food to survive, and food is obtained by foraging, therefore the animal needs to do *foraging behaviour. Needs can be distinguished from wants. Animals have *motivation to do (i.e. want to do) various types of behaviour, only some of which result in needs. Thus an animal may be motivated sexually, but sexual behaviour is not essential for survival of the individual. The animal does not therefore need sex in the same way as it needs food. Some scientists believe that everything that animals do is good for them in some way (meaning good for the propagation of their genes), but obviously some aspects of behaviour are more important than others in the long term. The relative importance of different activities can be estimated from studies of *demand and *resilience. Such studies tend to show that animals attach more importance to activities that result in essential needs, and least importance to *leisure activities.

negative reinforcer A *reinforcer that influences an animal to respond negatively to, and learn to avoid, the stimulus situation that is associated with the reinforcer. Examples include, loud noises, electric shock, or an attack from a conspecific or a predator.

neophilic Attracted to novel situations (*see* CURIOSITY).

neophobic Aversive to novel situations.

nerve A bundle of axons from various *neurons, serving to transmit information for relatively long distances within the body.

nest-building The behaviour involved in building a nest. A nest is a place where eggs are laid and hatched (especially in some insects, fish, reptiles, and birds), or (in mammals) where the young are born and nurtured.

The reptile ancestors of birds buried their eggs in the ground, where they took a long time to hatch. The present-day megapodes (Megapodiidae) of Australia and adjacent islands also bury their eggs in the ground, and depend upon the heat of the sun, or from decaying vegetation, to provide warmth during *incubation. Of these, the mallee fowl (*Leipoa ocellata*) is far more efficient than any known reptile in regulating the temperature around the eggs in its mound nest. This is done by scratching sand on or off the mound according to the ambient temperature.

Most birds sit on their eggs, thus providing warmth and protection from predators. They may do this in cavity nests, open nests, or roofed nests. The cavities may be excavated, as in woodpeckers (*Picidae*), but usually they are natural cavities, or those made by other animals. In these, nest-building is merely a matter of lining the nest with suitable materials.

Open nests may be made on the ground, or in trees or bushes. Ground-nesting birds usually begin their nest by making a circular scrape with the feet. They then line the hollow with various materials that help to insulate the eggs from the ground. Many birds, such as the Canada goose (*Branta canadensis*), then build up the surrounding rim of nest material. Ground-nesting birds usually have **precocial** young that run around and feed themselves from the day of *hatching. Birds that have **altricial** young, which are blind and helpless on hatching and depend upon their parents for food, usually build arboreal nests.

Arboreal nests built by large birds, such as herons (Ardeidae) and eagles (Accipitridae), contain large twigs and branches that are not easily displaced by wind. Medium-sized birds use twigs or grasses, sometimes adding mud to help bind the nest materials. Small birds use spider and insect silk to attach the nest to the substrate. Materials are usually inserted into the nest using thrusting and trembling movements. The nest cavity is formed, as in ground nests, by scraping movements of the feet.

Nests attached to vertical faces of cliffs or buildings are usually attached by using mud or saliva. Cave swiftlets (*Collocalia* sp.) have evolved an adhesive saliva, which is often mixed with plant materials.

Natural selection in different, but similar, environments has led to the convergent evolution of roofed nests. These may be made of woven or thatched grasses, as in weaverbirds (Ploceinae), or made of short heterogeneous plant materials bound together by spider or insect silk, as in sunbirds (Nectariniidae), or built of mud as in some swallows (Hirundinidae), or from a mortar made of sand and cow dung, as in the rufous ovenbird (*Furnarius rufus*). In compound nests, birds of the same species occupy separate compartments in the same nest mass. The compound nest of the sociable weaver (*Philetairus socius*), from the deserts of south-west Africa, may measure 1 by 7 metres and house 60 separate nest chambers. The birds live in their nests all the year round, and on cold winter nights several birds may sleep in one chamber, leaving some unoccupied. During the breeding season, the birds occupy separate chambers as pairs or families.

neuron A cell specialized for the transmission of information within the body. Each cell has a body containing a nucleus and a number of short branches called **dendrites**, and a single long protrusion called the **axon**. The dendrites connect with other neurons, and the axon conveys messages over relatively long distances within the body.

niche The role played by a species in the *community, in terms of its relationship both to other organisms and to the physical environment.

Thus a herbivore eats plant material and is usually preyed upon by carnivores. The species occupying a given niche varies from one part of the world to another. For example, a small herbivore niche is occupied by hares and rabbits (Leporidae) in northern temperate regions, the agouti (*Dasyprocta* sp.) and viscacha (*Lagostomus maximus*) in South America, the hyrax (Hyracoidea) and mouse deer (Tragulidae) in Africa, and by wallabies (*Wallabia* spp.) in Australia.

When animals of different species use the same resources or have certain preference or *tolerance ranges in common, **niche overlap** occurs. Niche overlap leads to *competition, especially when resources are in short supply. The *competitive exclusion principle states that two species with identical niches cannot live together in the same place at the same time when resources are limited. The corollary is that, if two species coexist, there must be ecological differences between them.

novelty responses Responses to unfamiliar stimuli. An animal's response to novelty depends partly upon the nature of the novel stimulus or situation, and partly upon the animal's internal state. When presented with a novel stimulus, a bird or mammal typically shows an *orienting response, a reflex turning of the head or body so that the eyes and ears are focused on the object. The animal then pays attention to the object and this is usually accompanied by a certain amount of autonomic *arousal. If a novel stimulus is presented repeatedly, *habituation occurs.

When released into a novel situation, such as a new cage, animals initially show signs of *fear and birds, especially, may remain motionless for a long time. This is often followed by *curiosity and *exploratory behaviour. The probability that a given novel situation will elicit exploration rather than fear, depends upon the animal's internal state. Animals that have recently had a frightening experience, or have been reared in isolation, are more likely to be wary of novel situations. Newborn animals investigate unfamiliar situations without signs of fear, but this boldness decreases with experience. As some situations become familiar, the response to novelty becomes increasingly fearful.

oestrous cycle A repeated pattern of periodic female sexual receptivity, in which bodily changes occur in response to fluctuations in the levels of circulating *hormones. Animals with an oestrous cycle are receptive only at the time of ovulation, in contrast to those with a *menstrual cycle, which are usually sexually receptive most of the time.

olfaction A type of *chemoreception in which low concentrations of airborne substances are detected by specially developed olfactory organs. Both taste and smell depend upon chemoreceptors. In both cases the chemicals are presented to the receptor in solution, and the distinction is difficult to justify in some animals, especially those that live in water. Nevertheless, in many animals there is a neurological distinction, in that some nerves are concerned with relaying olfactory messages, and others gustatory messages. In the blowfly (*Phormia regina*), for example, chemoreceptors in the antennae detect small quantities of airborne substances, and chemoreceptors in the tarsi (feet) are capable of detecting salt, sugar, and pure water. In vertebrates, the sense of taste is relayed via the facial (VII) and glossopharyngeal (IX) cranial nerves, while the sense of smell is transmitted by the olfactory nerve (I).

ontogeny Synonymous with *development, and concerned with the history of the individual from conception to death, and with the roles of *genetics, *maturation, and *learning in shaping the animal's life history.

Many behaviour patterns seem to be *innate, in the sense that they develop without example or practice. Thus, newborn domestic chicks (*Gallus gallus domesticus*) start pecking at small objects as soon as they are hatched. However, most scientists recognize that all behaviour is influenced to some extent by the animal's genetic make-up, and, at the same time, by the environmental conditions that exist during development. The extent to which the two influences, nature and nurture, determine the outcome varies greatly from species to species, and from activity to activity within a species. Behaviour that is inborn in the sense that it has high *heritability is, nevertheless, influenced during development. Thus the pecking of newly born chicks is somewhat indiscriminate and improves in selectivity and accuracy with experience.

Much animal behaviour is innate in the sense that it inevitably appears as part of an animal's repertoire under normal and natural circumstances. An activity can be innate in this usage, even though it is learned. For example, the juveniles of many species learn the characteristics of their own species members, and of their *habitat, at a particular stage of development. For this *imprinting to occur in the proper manner, the *parental care must follow its normal pattern. Thus, although genetic factors may be necessary for certain behaviour to develop, the species-characteristic processes of *maturation and *learning may be just as important. All aspects of ontogeny involve irreversible changes. Even learned responses, which may be subject to *extinction or *forgetting, nevertheless change the animal's psychological make-up for ever.

The development of behaviour is often characterized by specialized *juvenile behaviour. Like adult behaviour, this is subject to *natural selection. Thus, although some aspects of juvenile behaviour are precursors of adult behaviour, other aspects are specifically adapted to the survival of the young animal. For example the *alarm responses of herring gull (*Larus argentatus*) chicks are quite different from those of the parents. When alarmed, the chicks move a short distance from the nest and crouch silent and motionless among the vegetation, whereas the adults fly away uttering alarm calls. Sometimes, the lifestyle of the juvenile is quite different from that of the adult. Examples are the tadpoles of frogs (Anura) and the caterpillar larvae of butterflies (Lepidoptera). In other species, such as the wildebeest (*Connochaetes* sp.), the young animal runs with the herd within a few minutes of being born, and has a lifestyle that is almost identical to that of its parents.

Growth and behaviour development often go hand in hand, and many juveniles have some ability to 'catch up' if their developmental progress is interrupted by illness or food deprivation. For example, rats (*Rattus norvegicus*) fed a deficient diet between the ages of 9 and 12 weeks lose weight. When fed a normal diet again at 12 weeks they put on weight rapidly, and by 14 weeks are back at the weight that they would have been if they had not been deprived.

As the animal matures, new behaviour patterns appear, and this is especially true of *sexual behaviour, which develops under the influence of the sex *hormones. In addition to maturation, which does not require experience, some behaviour patterns develop with *practice. Repeated movements accelerate the development of behaviour. For example, newly hatched cuttlefish (*Sepia* sp.) take as long as 2 min to fixate and seize their prey. The time required diminishes rapidly with practice, and reaches 5 s in about ten trials. This improvement is the same whether or not the cuttlefish is hungry, and irrespective of its success in obtaining shrimps (*Corophium* spp.). However, the improvement does depend upon the experience of trying to catch shrimps and does not simply improve with time. Thus the improvement is not due to maturation or *learning. Unlike learning, practice depends upon experience, but not upon specific consequences of the behaviour.

Many animals have periods of development when they are especially sensitive to particular stimuli, especially those associated with their parents and siblings. During these sensitive phases of development certain types of learning, especially *imprinting, are facilitated. Examples include language learning in humans and *song learning in birds. In addition to immediate effects upon parent–offspring *relationships, imprinting can have marked effects upon the *social relationships of adults, and on *food selection and *habitat selection. The sexual preferences of many birds are influenced by early experience, a phenomenon called **sexual imprinting**. In ducks (Anatidae), domestic fowl (*Gallus gallus domesticus*), pigeons (*Columbia livia*), and finches (Fringillidae), individuals of one breed with a distinctive coloration can be reared by parents of a different breed and colour. When mature, such individuals prefer to mate with birds of their foster parents' colour, rather than their own colour. Hand-reared birds often become sexually imprinted upon people, and hand-reared mammals often develop special relationships with people.

operant conditioning A type of *conditioning (also called *instrumental conditioning), in which the animal is trained to make a particular response to obtain *reinforcement by a process called **shaping**. The experimenter places an animal, such as a pigeon (*Columbia livia*), in a suitable apparatus. For pigeons, the apparatus is a small cage equipped with a mechanism for delivering grain and a key (a small illuminated 'door' that operates a micro-switch when opened) at head height. Delivery of food is usually signalled by a small light that illuminates the grain. Pigeons soon learn to associate the light onset with the delivery of food, and to approach the grain and eat for a few seconds, after which the light goes off and the grain is removed. The next stage is to make food delivery contingent upon some aspect of the animal's behaviour. Initially the pigeon is rewarded for approaching the key, then for touching the key with its beak, and finally for pecking the key. Once the pigeon reaches this stage, the pigeon is said to have been shaped to peck the key, and key-pecking is automatically (via the micro-switch) followed by delivery of grain.

Many species of animal have been shaped in this way, although the apparatus must be appropriate to the animal concerned. For example, rats are usually required to press a lever to obtain reward. It used to be thought that any behaviour, of any species, could be shaped by well-timed delivery of any type of reward. This is now known not to be the case. Thus pigeons readily learn to peck a key, but have great difficulty in learning to press a lever with the foot. Pigeons will learn to peck a key to obtain a sex reward (a 10 second view of a mate), but they then start courting the key, as though it was their mate. Pigs (*Sus*) will learn to press a lever downwards with their nose to obtain food reward, but they then insist on trying to root up the lever, an activity that they would use to obtain food in the wild. It is evident that operant conditioning is heavily constrained by the animal's *innate response to stimuli associated with reward, be it food, water, sex, or *avoidance.

opportunism The tendency to exploit an opportunity, a favourable set of circumstances that is accessible to the animal. At the species level, this generally applies to the exploitation of new *habitats. To be able to make the most of such an opportunity, an animal must be able to produce many offspring with good dispersal abilities. An example is the colonization of parts of Australia by rabbits (*Oryctolagus cuniculus*). There are also many examples among *parasitic animals. Such species are capable of rapid reproduction when circumstances are favourable.

At the individual level, opportunism often takes the form of rapid exploitation of new sources of food. This is common amongst omnivores. Thus herring gulls (*Larus argentatus*) in the UK have been quick to exploit the food that is available at refuse dumps, and gull populations have grown in certain areas. For the individual, an opportunity can be defined in terms of the benefits that would accrue, at any instant, were the animal to decide to indulge in the appropriate activity. An **opportunity profile** is a graph of such instants plotted as a function of time. Opportunity profiles are features of environmental change. Thus herring gulls foraging at refuse dumps usually have a limited time to harvest the available food, after it is deposited and before it is covered over with inorganic matter.

optimal behaviour The best behaviour that an individual can perform in the given circumstances, in accordance with particular *optimality criteria. For example, crows (*Corvus*) hunting for whelks at low tide usually select the largest ones. They then hover over a rock and drop the whelk so that it breaks open, exposing the edible inside. The number of times a whelk has to be dropped in order to break is related to the height of the drop. The crows have to expend energy in flying up to drop a whelk, so what is the best height for the crow to fly? In other words, what is the optimal behaviour—to make many low flights, dropping the whelk each time, or to make few, higher flights? Given that the aim of the exercise (the optimality criterion) is to break open the whelk while expending the least amount of energy, it turns out that the optimal behaviour is for the crow to drop the whelk from a height of 5 m, in which case it may have to make about five attempts, performing a total amount of upward flying of about 25 m. For heights smaller or greater than 5 m, a larger amount of total upward flying is required.

optimality criteria The criteria in relation to which it is possible to determine which of a set of alternatives is the best. In animal behaviour studies, optimality criteria are usually framed in terms of *fitness. The ideal criterion would be **inclusive fitness** but this is normally impractical. In most studies, some short-term index of fitness is employed as the optimality criterion. For example, in studies of foraging behaviour, the notion of **profitability** is often the criterion used to judge the best foraging strategy. In other types of study, the concept of *utility is the optimality criterion.

Optimality criteria are important wherever there is a **trade-off** among the various costs and benefits of an activity. For example, a certain type of prey may be the most profitable for a foraging animal, but the time spent hunting that prey may also be time that the forager itself is exposed to predation. There is a trade-off between profitability and exposure time, in this case, and the optimality criteria must take into account both the energetic and the temporal aspects of foraging.

Optimality criteria may be employed to calculate the optimal behaviour strategy or optimal design (e.g. of a limb), but the optimal solution is not always a stable one. For example, there are advantages and disadvantages of living in groups, and it would seem to follow that there should be an optimal group size for a particular species. However, if there were a group of the optimal size, then it would pay a solitary individual to join the group, thus pushing the group above the optimal size. The optimal group is unstable, and a stable group will, in practice, tend to be larger than the optimum.

An *evolutionarily stable strategy (ESS), in particular, cannot be an optimal strategy because, by definition, it cannot be bettered by any feasible alternative strategy, provided sufficient members of the population adopt it. ESSs are found in cases where the best strategy for an individual depends upon the strategies adopted by other members of the population.

optomotor reflex A reflex turning of the eyes or head in response to the horizontal movement of environmental objects. For example, when a fly (*Eristalis* sp.) is placed inside a cylinder painted with vertical stripes, it shows a typical optomotor reflex, turning in the direction of the stripes when the cylinder is rotated. Such reflexes do not occur when the fly moves of its own accord, although the visual stimulation is similar. This suggests that some type of *reafference principle is involved.

orientation The positioning of the body, or parts of the body, in relation to the environment. To describe animal orientation accurately, a **frame of reference** is required. In the case of **positional orientation** body parts may be described in relation to an external frame of reference, as illustrated in Figure 7 (top right), the whole body in relation to an external frame (top middle), and body parts in relation to an internal frame of reference (top left). The external frame of reference is often the vertical (vertical bar), which many animals can sense by

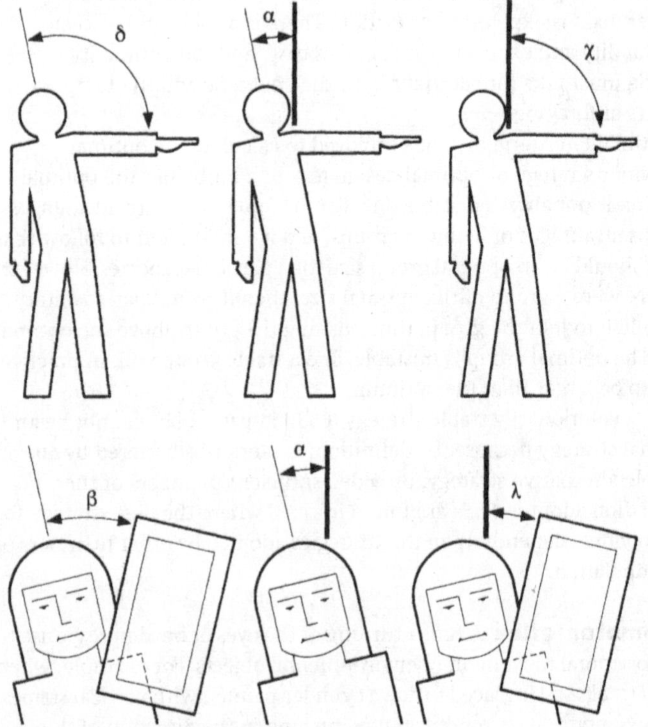

Fig. 7. Frames of reference in the orientation of body parts and of 'non-body' objects (the vertical bar indicates the direction of gravity).

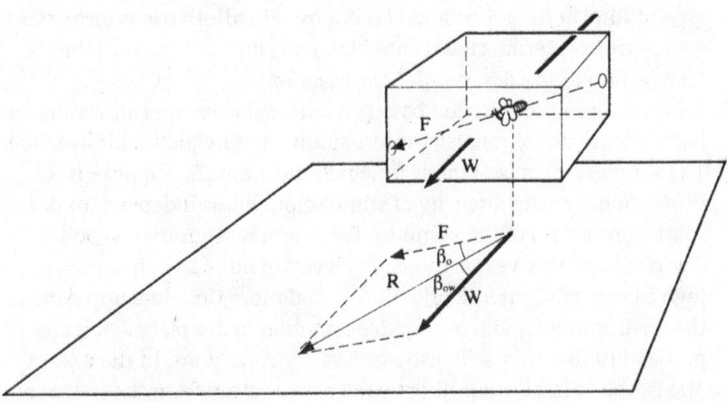

Fig. 8. The orientation of an animal in space. The flight path F of the animal is affected by the direction of the wind W, giving a resultant track R. The angles β_o and β_{ow} are the drift angle and the track angle, respectively.

means of *gravity receptors. Alternatively, orientation may be described in relation to an external non-body object (Figure 7, bottom row).

In the case of **course orientation**, the spatial control of translatory direction, the frame of reference may be some type of compass, such as a sun-compass or magnetic-compass, or a prominent landmark. In the case of flying animals, the situation is complicated by the fact that the animal's translation with respect to the ground is different from its translation with respect to the air, because the air moves in relation to the ground, as illustrated (*see* Figure 8) The airborne animal senses no wind, and sees only a ground pattern flowing in the opposite direction to its track. The angle between the track and the animal's heading direction is called the **drift angle**, as it depends upon the animal's drift due to the wind. The angle between the track and the wind direction is called the **track angle**.

Orientation in the absence of external cues is possible, in familiar surroundings, by means of *kinaesthetic senses. We can move in the dark, say towards a light switch, even with our ears covered, by means of a movement pattern consisting of a sequence of actions, that unrolls according to a learned program of motions. Many skilled actions, such as the golf swing, are controlled in this way. Each spatial manoeuvre is based upon the spatial situation left by the previous manoeuvre. This type of **idiothetic** information (as opposed to **allothetic** information gained from external cues) is used by many animals, such as the *navigation of the desert ant, *Cataglyphis Fortis*.

In the case of orientation based on external cues, we can distinguish between the use of intensity information and directional information. The simplest form is *kinesis in which the animal's response is proportional to the intensity of stimulation, but is independent of the spatial properties of the stimulus. For example, common woodlice (*Porcellio scaber*) are very active at low levels of humidity and less active at high humidity. Consequently, they spend more time in damp parts of the environment and their rapid locomotion in dry places increases the probability that they will discover less dry conditions. In the case of *taxis, the animal heads directly towards or away from the source of stimulation. The type of orientation that can be achieved in a given situation depends jointly upon the nature of the external cues and the sensory equipment of the animal. An animal with one sensor that is sensitive to stimulation strength only is limited to successive measurements of stimulus strength in different localities. If the external cues are inherently directional, then a single receptor that is shielded on one side can provide directional information. Thus a shielded photoreceptor is useful in this respect, but a shielded chemoreceptor is of no advantage, because chemical stimuli are not inherently directional. With two receptors, simultaneous comparison can be used to detect gradients, and with many receptors arranged in the form of a rastor (a row or mosaic arrangement) more sophisticated types of orientation can be achieved. Examples of rastors are the lens eyes of vertebrates and the compound eyes of arthropods.

Spatial orientation is often achieved by a combination of methods. For example, some moths (Lepidoptera) are attracted to females as a result of the airborne pheromone released by the female. The scent is windborne, and the flying moth must orientate with respect to the wind. Flying animals usually use visual cues to monitor their progress with respect to the ground. The flight path of the animal is affected by wind direction, giving a resultant track. The track angle changes with scent concentration. When the scent is absent, the animal flies backward and forward without progressing upwind. When scent is detected in the wind, the track angle increases and the animal zigzags upwind. The changes in direction are related to the borders of the scent trail. When the scent concentration drops at the edge of the scent plume, the animal turns in the direction opposite to that of the previous turn, using idiothetic information. Thus the flying moth uses a combination of visual, amenotaxis (wind), and idiothetic orientation mechanisms in searching for a mate.

Course orientation is important for *navigation. Many navigating animals make use of a magnetic, stellar, or solar compass to maintain directional stability. Homing ants (Hymenoptera), fish, and birds are known to use the sun as a compass. The animal maintains a course with a particular angle to the sun. At the outset, the angle is that which will provide the desired direction, but as time goes by, the sun moves. The animal compensates for the sun's movement by altering the angle in accordance with its internal *biological clock. This time-compensation ensures that a straight course is maintained.

orienting response The response made when an animal is suddenly presented with stimuli. It quickly turns its body or head so that its eyes and ears can be brought to bear upon the source of stimulation. The orienting response was first described by *Pavlov and is an involuntary *reflex. It has been called a 'what is it?' response that serves to focus the animal's *attention on the stimulus, and it is usually accompanied by some kind of autonomic *arousal. The behaviour that follows the orienting response depends upon the nature of the stimulus. A potentially dangerous stimulus may elicit *defensive behaviour, *escape, or *aggression. A response to *novelty may include *curiosity and *exploratory behaviour. Repeated presentation of a stimulus leads to *habituation of the orienting response.

orthokinesis A form of *orientation in which the animal's speed of locomotion is proportional to the intensity of stimulation, and is independent of the spatial properties of the stimulus. For example, common woodlice (*Porcellio scaber*) are very active at low levels of humidity and less active at high humidity. Consequently, they spend more time in damp parts of the environment and their rapid locomotion in dry places increases the probability that they will discover less dry conditions. Woodlice tend to form *aggregations in damp places beneath rocks and fallen logs, and their selection of this *habitat is based entirely upon orthokinesis.

P

pain An aspect of *emotion. The problem of assessing pain in animals is that it can only be done by analogy with ourselves, and this approach is open to the objection that it involves *anthropomorphism. However, since pain cannot be defined in an objective scientific manner, there is little option but to regard pain in animals as being similar to pain in humans.

Pain in humans is philosophically controversial. If a person says 'I am in pain' we cannot argue that they may be mistaken in the way that we could if they said 'It is raining'. There is no way we can verify that a person feels pain (or is telling a lie), even by physiological measurement. Normally, when a person says they are in pain, there are characteristic physiological indices that can be monitored. But the absence of physiological symptoms of pain does not enable us to conclude that the person is not in pain even though they say they are.

Some animals, especially mammals, have similar behavioural and physiological reactions to stimuli that would induce pain in us, and we would normally assume that the presence of such reactions indicates that the animal is in pain. But this is an assumption based on analogy, not a scientifically verifiable conclusion. The problem is confounded by the fact that non-mammalian animals, with nervous systems very different from ours, show characteristic behavioural reactions to pain stimuli, but their physiological reactions are somewhat different from ours. In the interests of animal *welfare, we can give animals the benefit of the doubt, but we have no way of finding out whether their inner experiences of pain are similar to ours, or whether they feel pain at all. Moreover, the concept of pain is not really necessary to explain animals' *avoidance and *escape reactions to so-called pain stimuli.

parasitism An interaction between members of different species, in which one (the parasite) exploits the other (the host), but generally does not kill it. Many parasites are fully dependent upon their host at some stage of their life cycle, and *natural selection therefore exerts a strong pressure for them to become highly adapted to their *niche. At the same time, many host species have evolved *defensive measures, so there is often an evolutionary 'arms race' between parasite and host.

A parasite that is dependent upon its host throughout its life is called an **obligate** parasite. These may be **endoparasitic**, living inside the host, as do tapeworms (*Cestoda*), or they may be **ectoparasites**, such as feather lice (*Mallophaga*) that live on the bodies of birds and feed upon blood and feathers.

Facultative parasites can live and complete their life cycles as free animals, but resort to parasitism when conditions are favourable. In many countries cats and dogs live as feral animals, and intermittently parasitize themselves on human beings.

Social parasites pass themselves off as members of another species, and are looked after by members of the host species. Social parasitism in common in insects. Brood parasitism is a form of social parasitism, in which birds of one species (e.g. the European cuckoo, *Cuculus canorus*) lay their eggs in the nest of another species. The host species then rears the parasitic species at its own cost.

parental behaviour Behaviour that aids the development of offspring. Rudimentary forms of parental behaviour may involve simply placing the eggs in a sheltered or secure location, and leaving them to develop unaided. This is common among insects and some amphibians and fishes. Some, such as the digger wasp (*Bembex rostrata*), not only place the eggs in a sheltered spot (a burrow in the ground), but also provide food for the emergent larvae. The female digger wasp provisions her nest with the carcass of a paralysed insect.

Among fishes and amphibians there is a spectrum of parental behaviour, from external fertilization and virtual abandonment of the eggs, through various types of egg-guarding, to true viviparity. This occurs in sharks (Selachii) and a number of species of bony fishes (Teleostei). In the toad *Nectophrynoides* spp., the young develop completely within the reproductive tract of the female, the tadpoles gaining their nutrients through the blood vessels of the tail, which co-circulate with those of the oviduct. This arrangement is the equivalent of the mammalian placenta.

Amongst reptiles, internal fertilization is almost universal, and various forms of viviparity have evolved. Parental behaviour is highly developed in crocodiles. Elaborate nests may be built and the nest is usually guarded. The female of the Nile crocodile (*Crocodilus niloticus*) transports the young in her mouth to the edge of the water as soon as they are hatched. She stays in the vicinity of her young for many weeks during their early development.

Viviparity has not evolved in birds, but care of the young is often complex, and may last until the young reach maturity. It may involve *nest-building, *incubation, nest guarding, and feeding the young, often by both parents. The degree of *parental investment is considerable, and consequently pair bonds are strong and *monogamy common.

Among mammals, the picture is completely different, because the females are specially equipped (with mammary glands) to feed the young. Pair bonds tend to be temporary, and *monogamy uncommon, although it does occur in some primates. The *maternal behaviour that is characteristic of mammals, often involves *imprinting by the young, and *teaching of the young. The role played by males is often minimal.

parental care A form of *altruism in which, by spending time and energy in aiding its offspring, the parent is increasing the *fitness of the offspring to the detriment of its individual fitness; it is favouring current offspring at the expense of possible future offspring. The degree of parental care varies considerably from species to species. It depends upon the number of offspring normally produced in a lifetime, the type of *mating system involved, and the aid given to the offspring by animals other than the parents.

in a tree, do not avoid the visual cliff. In fact, when they approach the edge of the cliff they make a little jump, which is what they would do in nature when exiting the nest.

Studies of *imprinting and of abnormal rearing environments, show that *learning plays an important part in some aspects of perception, *song learning in birds, for example. When a kitten (Felidae) raised in a vertically striped environment was put together with one raised in a horizontally striped environment, there were clear differences in their behaviour. If a rod was held vertically and moved about, the vertically raised kitten would play with it, but the other would ignore it. When a rod was moved about in the horizontal plane, the horizontal kitten played with it, but the vertical kitten ignored it. Thus rearing conditions can have effects upon perceptual development.

perceptual organization The organization of sensory data to provide a framework for recognition of objects in the external world. For example, an animal that uses *vision to guide its way around the world must extract information about the size and distance of objects. This can be done in a number of ways.

(1) **Motion parallax** If the animal moves its head, the images of nearby objects are swept across the retina faster than those of far away objects. Many animals use this cue. Thus, when locusts (*Orthoptera*) prepare to jump across a gap, they move their heads from side to side a few times, before leaping.

(2) **Retinal disparity** Since each eye views the world from a different viewpoint, the images of objects at different distances fall on different parts of the two retinas. This retinal disparity can be used to gauge the distances of objects. The praying mantis (*Mantis religiosa*), and many birds and mammals, use retinal disparity as a cue to depth.

(3) **Looming** When an object approaches, its retinal image increases in size. Many animals, including crabs (Brachyura), frogs (Anura), chickens (*Gallus gallus domesticus*), kittens (Felidae), and monkeys (Simiae), respond to this stimulus by running away.

(4) **Shadows** Concave objects cast a shadow in the region nearest to the light source, while convex objects cast a shadow in the region furthest from the light source. In nature the light source (the sun) is above the animal, so to discriminate between concave and

convex objects it is necessary only to look at the direction of the shadow. Many animals use this cue in object recognition.

In many animals, information from different senses is coordinated. When suddenly presented with a stimulus, the animal performs a reflex *orienting response. It quickly turns its body or head so that its eyes and ears can be brought to bear upon the source of stimulation. Not only are the senses behaviourally coordinated, they are often internally coordinated. For example, objects identified by auditory location are then visually fixated.

In addition to perceptual organization of the external world, animals have to cope with the information from the internal *sense organs that detect changes in their own body, sometimes called the kinaesthetic *senses. For example, most species, from jellyfish (*Aurelia* sp.) to humans, have organs of balance that enable them to maintain an upright *orientation.

Eye movements in mammals cause images to move on the retina, but the animal does not interpret this as movement of the outside world. If the retinal image moves as a result of the animal's own movement, then this is compensated for via a *reafference mechanism. The result is that the animal perceives no movement under these conditions. Other mechanisms serve a similar role in birds and in some invertebrates. Thus, the perceptual organization of internal information enables animals to distinguish between changes caused by their own bodily movements, and changes emanating from the outside world.

p

peripheral nervous system That part of the nervous system that does not include the *central nervous system. In invertebrates, the peripheral nervous system includes the network of nerve fibres emanating from the anterior ganglia and nerve chords, while in vertebrates it includes the nerves of the somatic nervous system, supplying the muscles, and the autonomic nervous system, supplying the intestines and other interior organs.

pheromones A chemical complex that is released into the environment by an organism and causes a specific behavioural or physiological response in a receiving organism of the same species. Pheromones are thus a means of chemical *communication. For example, the female silkworm moth (*Bombyx mori*) releases a polyalcohol, known as bombykol, which can attract males over distances of several kilometres. The action of the pheromone is very specific, and a single molecule is sufficient to trigger the response. The male flies upwind when it detects bombykol molecules, and when he finds the female he copulates with her.

Pheromones are also released by aquatic animals, particularly as part of their *alarm reactions. In vertebrates, pheromones are detected via the *olfactory system. They can be important as *sexual attractants, in *scent marking, and in *territorial behaviour.

photoreceptor A type of *sensory receptor that is sensitive to light (*see* *vision).

phototaxis *orientation to light involving *taxes (movements in relation to the light source).

play An aspect of *juvenile behaviour, in which the (usually) young animal spends time in apparently pointless activity, such as friendly fighting, sex without coition, hunting without prey, etc. Play is often accompanied by a characteristic *facial expression and characteristically energetic movement. It is a type of *leisure activity, in that it disappears from an animal's repertoire when demands upon the animal's time are very severe. Although play seems to be functionless, it may be a type of rehearsal or *practice for activities that will become important later in life. It may also be a form of exploration of both the physical and the social environment.

poison avoidance An aspect of *learning by which animals rapidly learn to avoid poisoned food. Rodents are notorious for their bait shyness. Rats (*Rattus norvegicus*) that eat a small amount of poisoned food and survive will never touch that type of food, again. When a rat comes across a novel food, it initially eats only a very small amount. The intervals between meals are sufficiently long to enable the rat to assess the consequences of eating a new food. When a rat eats, some of the food passes rapidly through the stomach into the small intestine, bypassing the previous meal that is still in the stomach. In this way the rat can test small quantities of the food that it has eaten most recently. Rats show considerable *stimulus relevance, readily associating the novel taste or smell of food with internal physiological consequences, and not associating them with other consequences. Even if the consequences of eating are delayed by 24 h, the rat associates deleterious consequences, such as sickness, with the food it has eaten, and not with other behaviour that has occurred during the interval. A single experience of a poisoned food is sufficient to establish avoidance of that type of food. Similar phenomena occur in birds, although they usually form avoidance associations with the visual properties of the food. The mechanisms of poison avoidance have a close affinity with those of *food selection in general.

polygamy A type of mating system in which each adult may mate with more than one member of the opposite sex. In **promiscuous** species, in which both males and females routinely mate with more than one member of the opposite sex, there are no pair bonds and *parental care is minimal. In the case of **polyandry**, the females mate with a number of males, while in **polygyny** males mate with many females and father a large number of offspring. A successful polygynous male has little time for his offspring. Either they must be self-sufficient, or the female must be able to care for them unaided. Polygyny will usually result in a few successful males and a large number of unsuccessful males. The harem may then become the basic feature of the social structure, as in many antelopes (Bovidae) and deer (Cervidae).

positive reinforcer A *reinforcer that encourages the animal to respond positively, and learn about, the stimulus situation that it associates with the reinforcer. Examples include food, water, or the sight of a mate.

practice An aspect of *ontogeny in which repeated movements accelerate the development of behaviour. For example, newly hatched cuttlefish (*Sepia* sp.) take as long as 2 min to fixate and seize their prey. The time required diminishes rapidly with practice and reaches 5 s in about ten trials. This improvement is the same whether or not the cuttlefish is hungry, and irrespective of its success in obtaining shrimps (*Corophium* spp.). However, the improvement does depend upon the experience of trying to catch shrimps and does not simply improve with time. Thus the improvement is not due to *maturation or *learning, but simply to practice. Practice depends upon experience, but not upon specific consequences of the behaviour, in contrast to learning.

precocial young The young of a precocial (usually bird) species, in which hatching occurs late in *ontogeny. In contrast to *altricial species, in which hatching occurs early, the precocial bird is still inside the egg when the first feathers appear. Precocial young can usually move around and feed themselves just after hatching.

predation A relationship between animals, in which one, the predator, benefits while the other, the prey, is affected adversely. This situation is in contrast to cases in which the relationship is symmetrical, as in *competition, where both participants are affected adversely. Predation implies that the prey is killed, and usually eaten, by the predator. It thus differs from *parasitism, in which one animal exploits another, but usually does not kill it.

Although predation usually occurs between members of different species, it may take the form of *cannibalism, in which an animal kills and eats a member of its own species. For example, herring gulls (*Larus argentatus*) prey upon the eggs and chicks of their neighbours during the breeding season.

predator avoidance An aspect of *defence designed to avoid predation. Much of this avoidance behaviour is neither wholly innate nor wholly learned. Many animals have species-specific defence reactions. For example, some may struggle to escape when caught by a predator, while other feign *death. New avoidance behaviour is more quickly learned if it is similar to the species-specific defence behaviour of the animal concerned.

Defence against predators may be primary or secondary. Primary defences operate regardless of whether or not there is a predator in the vicinity. They reduce the probability that a predator will encounter an animal. They include *camouflage and *mimicry, group living and some forms of *social interactions, and *symbiosis.

Secondary defences operate only after an animal has detected a predator. They increase the chances that the animal will *escape from the encounter. They include withdrawal, flight (escape), bluff, death feigning, deflection of attack, and retaliation. Bluff takes the form of *deimatic display designed to scare off a predator. For example, the caterpillar of the hawkmoth (*Leucorampha* sp.) normally rests upside down beneath a branch or leaf. When disturbed it raises and inflates its head, the ventral surface of which has conspicuous eye-like marks, and the general patterning of which resembles the head of a snake. Many moths and butterflies have eye-spots on their wings, which they reveal suddenly when disturbed, with the possible effect of frightening the predator.

predatory behaviour The behaviour by means of which an animal of one species, the predator, kills and eats a member of another species, the prey. The *motivation for predatory behaviour is usually *hunger, but this is not always so. In most animals, the speed and efficiency of prey capture increase with hunger, but in the praying mantis (*Hierodula crassa*) and the jumping spider (*Epiblemum scenicum*) the stereotyped movements of prey capture remain unaffected by hunger.

Many predatory species obtain food that they do not eat themselves, but give to their young. Usually, obtaining food for the young is under the control of stimuli from the young, rather than the hunger of the parents. Birds of many species adjust their parental foraging to the size of their brood. The begging behaviour of the young is the prime stimulus by means of which the parents adjust to their food requirements. The predatory behaviour itself may also be different when it is for the benefit of the young. Thus domestic cats (*Felis catus*) and cheetah (*Acinonyx jubatus*) usually kill their prey soon after capture, but when it is for the young they carry live prey to the litter and release it in the presence of the kittens. Some birds eat the food themselves and then regurgitate it to their young, while others select smaller prey items than those that they would eat themselves.

Many birds and mammals kill more than they eat and store the surplus. Ravens (*Corvus corax*) are more likely to store food when they are hungry, whereas shrikes (*Lanius* sp.) store more when they are satiated. Ravens store prey at the time that they are feeding their young. Thus ravens store in response to demand, whereas shrikes store in relation to the abundance of available prey. The relationship between *hunger and *hoarding is complex, differs from species to species, and serves different functions in different species.

Choice of prey is often dictated by availability, but many predators may concentrate on one prey and then suddenly switch to another. Thus redshank (*Tringa totanus*) when feeding upon the marine polychaete, *Nereis* sp., select large worms whenever possible and pass over small ones. If the shrimp *Corophium* spp. is available, these are taken preferentially, even though the redshank would gain much more energy by sticking to worms. This aspect of *food selection is almost certainly related to the particular nutritional requirements of the predator.

Many predators have *rhythms of predatory behaviour. These may be related to the availability of prey, or the rhythms may be endogenous, persisting independently of environmental factors. The northern shrike (*Lanius excubitor*) usually feeds upon insects and lizards, but in the autumn the lizard content of the diet drops to zero, even though the lizards remain plentiful. Many predators have a marked *circadian rhythm. Thus domestic cats have a pattern of nocturnal hunting that is independent of their food supply, but matches the availability of their natural prey.

Predators obtain their prey by *foraging, which may include *searching and *hunting, or may be a matter of lying in wait for prey to ambush. Some predators make use of *camouflage, while others use aggressive *mimicry, passing themselves off as a harmless creature. For example, the sabre-toothed blenny (*Aspidontus taeniatus*) closely resembles the cleaner wrasse (*Labroides dimidiatus*), both in its colouration and its behaviour. Large fish permit the cleaner wrasse to approach and remove parasites from their body surface and the inside of the mouth. The relationship is symbiotic, because the wrasse benefits by obtaining food and the host fish benefits by having parasites removed. Sometimes, the blenny is mistaken for the cleaner wrasse, in which case the mimic approaches the host fish, bites a piece from a fin, and escapes.

preening A form of *grooming behaviour performed by birds as part of feather maintenance. It is often seen in conjunction with other *comfort behaviour, such as bathing, dusting, and sunning.

Preening consists in the cleaning and arrangement of feathers, in response to dirt or wetness. Attention is given to individual feathers as well as the arrangement of feathers as a whole. Many birds have a preen gland near the tail that secretes an oily substance. The bird stimulates the flow of oil with its bill and then applies it to the feathers with stroking and quivering movements. Exposure to sunlight increases the vitamin D content of preen oil, which is sometimes ingested. The preen oil helps to keep the feathers supple and water resistant.

prey recognition The *discrimination of prey from non-prey. Most predators encounter a large number of different prey species that they have to discriminate from non-prey. The most commonly used cues are size, movement, and shape. Predators that have a choice of prey differing in size usually take the larger, as this is the most energy-efficient strategy. However, as size increases there usually comes a point beyond which the stimulus is no longer regarded as prey. Thus toads (*Bufo bufo*) respond positively to prey items within a specific size range, but actively avoid larger stimuli. The toad is able to judge the absolute size of the prey item by taking into account both the size of the image on the retina and the distance to the object. This is a matter of *perceptual organization.

Movement of the prey-stimulus is essential for prey recognition in some species. The cuttlefish (*Sepia officinalis*) will normally only attack prawns that are moving, but if a prawn is taken from a cuttlefish that has just captured and paralysed it, then it will immediately be attacked again, even though it is motionless. The cuttlefish is now using an olfactory cue.

Prey recognition by shape is a complex matter, but a number of predators are known to use bilateral symmetry as a cue. Most living creatures are bilaterally symmetrical, whereas inanimate objects are not. Toads show prey-catching behaviour towards small, elongated objects, but avoid snake-like shapes with one end raised.

problem solving The ability to reach a goal that seems initially unattainable. In the study of problem solving in animals, the problems are usually set by humans, but certain types of problem may occur in nature, particularly detour problems, and problems associated with harvesting food efficiently.

The ability to solve problems is often taken as a sign of *intelligent behaviour, but it is not possible to devise a fair intelligence test for an animal. Not only are some species good at some types of problem solving, and other species other types, but the performance of animals depends very much upon the testing apparatus. For example, rats (*Rattus norvegicus*) perform much better at problems involving visual discrimination, if they are required to jump at the symbol of their choice, instead of running at it. It is not possible to find the best testing apparatus for a given species, because not all possibilities can be tried. It is not fair to compare animals tested in different circumstances, nor is it fair to test animals of different species in the same apparatus.

The ability to solve problems is sometimes taken as an indication of *cognition or *insight. For example, field studies show that chimpanzees (*Pan* sp.) sometimes take a roundabout route to a destination, suggesting that they have a mental picture of the spatial relationships of objects within the environment. Such suggestions have proved to be controversial, and the question of whether animals can really think remains an open question.

An important problem, from a theoretical point of view, is the transitive inference problem. If A is bigger than B, and B is bigger than C, is A bigger or smaller than C? This problem can be presented in many forms. It is important, because transitive inference is the basis of *rationality. Young children, monkeys (Simiae), and even pigeons (*Columbia livia*), can solve transitive inference problems, indicating some innate ability to order things rationally. However, the way the problem is presented, whether to children, monkeys, or pigeons, makes a large difference to the outcome. Thus animal problem solving is, in general, a difficult area to investigate. Nevertheless, animals sometimes manage to solve seemingly intractable problems. Sheep (*Ovis*) in northern England have been observed to lie down and roll over 3 metre-wide cattle grids to obtain access to vegetable gardens.

procedural knowledge *Knowledge that is part of a procedure, and not applicable to other situations. Thus knowing how to ride a bicycle is procedural knowledge that cannot be passed to another person or used for other purposes.

pseudoconditioning An aspect of a *conditioning situation in which the unconditional response (UCR) may come to be elicited by stimuli other than the unconditional stimulus (UCS) (as in normal conditioning), even though there is no contingent relationship between them. A possible explanation is *generalization to stimuli similar to the UCS. It can also happen that exposure to some kind of shock can alter the animal's *motivational state so that it comes to avoid any external stimulus.

punishment Negative *reinforcement, usually used to promote *learning and not intended to allocate blame. However, punishment, such as administration of electric shock, may result in *fear, anxiety, and *stress, which may interfere with learning. Mild negative reinforcement, such as withholding a reward, or removal of the animal from the situation, has often been found to be more effective in promoting learning.

purposive behaviour Behaviour that has the appearance of being directed towards a *goal. Not all apparently purposive behaviour is truly goal-directed in the sense of being *intentional. True intention behaviour, that is carried out by reference to an explicit representation of a *goal, is difficult to identify in animals. *Deceitful behaviour is sometimes taken as evidence of true intention, but the issue remains controversial.

A second meaning of purposive behaviour is behaviour that is designed for a purpose. The normal biological terminology here is that behaviour may be designed (by natural selection) to serve some **function**. Thus the function (purpose) of a wing is to propel an animal through the air, meaning that the wing is designed to propel an animal through the air.

R

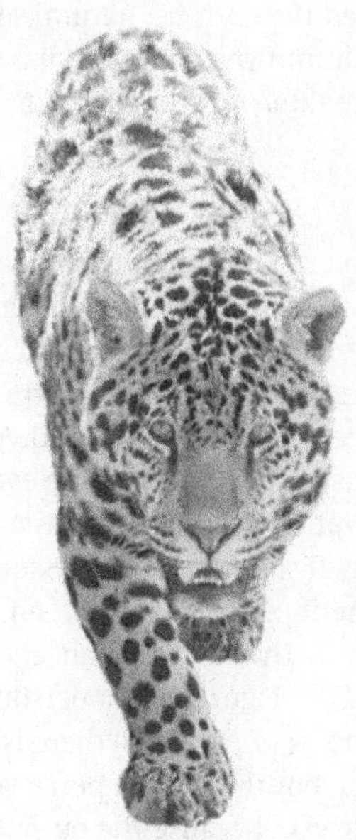

rationality An aspect of *decision-making. Rational decision-making is self-consistent, a property that is usually called transitivity-of-choice. Suppose an animal can choose between options A, B, and C. If A is preferred to B, and B is preferred to C, then it is rational to expect that A will be preferred to C. We can now write a consistent order of preference A>B>C. These relationships among A, B, and C are said to be transitive. If A>B and B>C, but A is not preferred to C, then the decision to choose C above A is irrational, and the relationships among A, B, and C are intransitive.

Rationality does not necessarily imply the use of reason, when the term is used as part of the study of *animal economics. The evidence suggests that animals do make transitive choices, and this implies that something is maximized in the decision-making process.

The name given to the quantity that is maximized by the process of rational decision-making is *utility. This term is also used in human economics, but biologists make a distinction between the quantity maximized by the individual animal, i.e. utility, and the quantity that should be maximized (or minimized) to attain maximum *fitness. This quantity is usually called the cost, i.e. natural selection favours those choices that incur minimum cost. An animal is considered to be well adapted if its decision-making is rational (i.e. maximizes utility) and also minimizes cost.

reafference An aspect of *orientation by means of which the animal is able to distinguish between exogenous changes in the environment and changes that result from the animal's own behaviour. For example, when we move our eyes the image on the retina moves, but we do not perceive this as movement of our visual world. According to reafference theory, the (efferent) instruction to move the eyes is accompanied by an efference copy (or output copy) that represents the movement of the image that would normally be expected to ensue from movement of the eye. The actual movement of the image is then compared with the output copy in such a way that the two cancel each other, and no movement is perceived (*see* Figure 9). If a person is prevented from moving the eye (by temporary paralysis) there is no actual movement of the image on the retina, but the person perceives movement in attempting to move the eye, because the output copy is not cancelled.

It is important for animals to be able to distinguish between changes in the world that are caused by their own behaviour, and changes that are independent of their behaviour. Thus a bird that lands on a branch bends the branch at the same time that the branch is being bent by the wind. In order for the bird to keep its balance, it must be able to

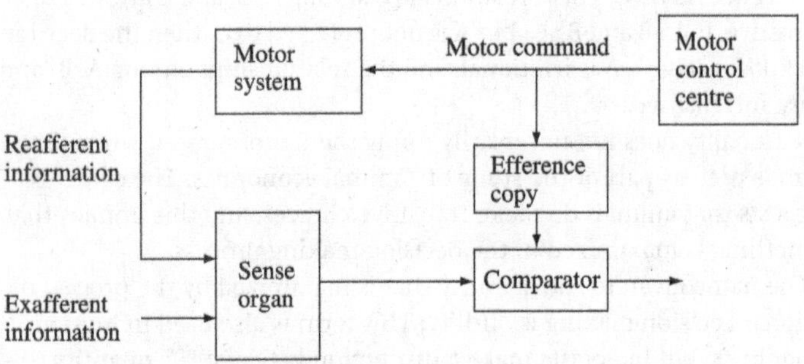

Fig. 9. Outline of a basic reafference system.

anticipate the deformation of the branch that will result from the impact of landing on it.

reciprocal altruism An aspect of *altruism in which one animal helps another at some cost to itself in the 'expectation' that it will be the a recipient of altruism at a later date. The problem with the evolution of this kind of altruism is that individuals who cheat by receiving but never giving would be at an advantage.

It is possible that cheating could be countered if individuals were altruistic only toward other individuals that were likely to reciprocate. For example, when a female olive baboon (*Papio anubis*) comes into oestrus, a male forms a consort relationship with her. He follows her around, waiting for an opportunity to mate, and guards the female from the attentions of other males. However, a rival male may sometimes solicit the help of a third male in an attempt to gain access to the female. While the solicited male challenges the consort male to a fight, the rival male gains access to the female. Those males that most often give aid of this type, also most frequently receive aid. Thus reciprocal altruism can be demonstrated in this situation.

recognition A process of *perception, by means of which the animal identifies objects in the external environment. For example, herring gulls (*Larus argentatus*) are able to identify eggs that have rolled out of the nest. The shape, size, background colour, and markings are all important in identifying the egg as an egg-to-be-retrieved, rather than a foreign object. It has been demonstrated that these features are additive in their contribution to egg recognition (i.e. the *law of heterogeneous summation is obeyed in this case).

Recognition of prey and predators is largely *innate, and involves species-specific *sign stimuli. However, many animals can be trained to recognize abstract stimuli. Thus pigeons (*Columbia livia*) can be trained to choose one of two stimuli that matches a sample stimulus, or to choose one of two stimuli that differs from a sample. They can learn to recognize shapes regardless of size or colour, and have been shown to remember some 700 arbitrary shapes, and to recognize them when tested some seven months later.

redirected behaviour Behaviour directed at external stimuli that are irrelevant to the current situation. For example, a thirsty pigeon (*Columbia livia*) that is denied access to water at an accustomed place, or time, may show drinking movements directed at smooth surfaces of shiny pebbles. In winter flocks of small birds, when one bird is supplanted at a food source by a superior, it may attack an inferior bird instead of retaliating.

reflexes The most simple form of reaction to stimulation. Stimuli, such as the sudden change of tension in a muscle, a sudden change in the level of illumination, or a touch on some part of the body, induce an automatic, involuntary, and stereotyped response. The *sense organs involved, such as muscle spindles or light detectors, are usually linked to specific reflex mechanisms. They send messages to the central nervous system, and instructions are then sent directly to the effector organs, usually muscles or glands. The reflex response is not entirely automatic, because there is generally a possibility of interference from within the central nervous system. There may be inhibition from other incompatible reflexes, or modifying influences from the *brain itself.

Reflexes are involved in the *coordination of limb movements, and in all aspects of *locomotion. In some cases, such as the startle reflex, the whole animal is involved. This is especially true of the *escape reflexes of many invertebrates.

Reflexes may be *innate or the result of *learning. Thus many new reflexes are established as the result of *classical conditioning.

regulators Animals that maintain their bodily functions in a condition that is relatively independent of environmental fluctuations, in contrast to *conformers that allow their internal environment to be influenced by the environment. Some degree of internal regulation was a prerequisite for the invasion of fresh water and of dry land.

reinforcement The process by which the tendency to perform certain responses becomes strengthened as a result of *learning. Reinforcement is not absolutely necessary for all types of learning, but normally an animal learns when its behaviour has novel consequences, and these consequences are said to reinforce the behaviour. Thus, if a hungry animal performs a particular activity and obtains food for the first time, then it will start to learn to perform this activity again, to obtain food. The activity occurs in the presence of certain stimuli. These become associated with the reinforcement, and the animal starts to respond to these stimuli as if they were food. Thus the reinforcement facilitates a kind of stimulus substitution.

A reinforcer that encourages an animal to approach the associated stimuli is generally called a *positive reinforcer. If the reinforcer leads the animal to avoid the situation in future, then it is a *negative reinforcer. Thus an animal may learn to fear a situation as a result of experiencing pain or stress there. They subsequently may show *avoidance behaviour when they next encounter the situation. Thus situations in which the animal encounters fear-inducing stimuli may induce fear by association, without the animal experiencing pain.

reinforcement schedules An aspect of *operant conditioning, in which the animal makes repeated responses to obtain a reward. Thus, a rat (*Rattus norvegicus*) may be required to press a bar a number of times to obtain a food reward. Instead of rewarding the rat for every response, it may be rewarded for every *n*th response, so that there is a fixed ratio between the number of responses and the number of rewards. This procedure is called a fixed-ratio reward schedule. Other common reinforcement schedules are variable-ratio, fixed-interval, and variable-interval. On an interval schedule, reward is given at intervals of time specified by the experimenter. The animal is rewarded for the first response made after the given interval has elapsed. Different schedules of reinforcement have been found to have different effects upon the animal's performance. For example, a variable-interval schedule produces a very uniform rate of responding and provides a good benchmark against which to test the effects on behaviour of various factors, such as reward size or reward quality.

reinforcer A consequence of behaviour that supports *learning. It may take the form of reward (positive reinforcer) or *punishment (negative reinforcer).

releaser An alternative term for *sign stimulus.

representations An aspect of *cognition in which it is assumed that *knowledge is represented in the nervous system of animals. There are two basic types of representation: implicit and explicit. **Implicit representation** is the same as information that is available as part of a fixed procedure. For example, to know how to touch my nose with my eyes closed, I have to know, amongst other things, the length of my forearm. This information is represented implicitly in my kinaesthetic system. The representation changes as I grow, and can be changed by *learning. Similarly, to ride a bicycle, I have to learn to coordinate numerous kinaesthetic systems to do with steering and balance. I can then say that I know how to ride a bicycle, but this type of knowledge is tied to the procedure of riding a bicycle. For this reason, the kind of knowledge that depends upon implicit representations is sometimes called procedural knowledge.

We now come to **explicit representations**. By 'explicit' it is meant that the information is made obvious in a physical manner, and is not simply part of a procedure. If a representation is to be explicit, then there has to be a physically identifiable bearer of the information (the token) and, additionally, something that can be identified as the user of the information. For example, many motor cars have icons on the dash that provide information that can be used by a competent user. The icons are tokens that indicate overheating, low oil pressure, etc. The competent user can recognize these and act accordingly. Whether non-human animals can have explicit representations is a matter of controversy.

reproductive behaviour All behaviour that has to do with reproduction. It includes *courtship, *mate selection, *sexual behaviour, and *parental behaviour.

reproductive isolation An aspect of *evolution in which various *isolating mechanisms may prevent interbreeding and gene exchange between closely related populations. Any inherited difference between male and female that prevents the production of reproductively viable offspring will lead to reproductive isolation, so that the two populations evolve into different species.

reproductive success An aspect of *fitness that is sometimes equated with Darwinian fitness, in that it is a measure of an animal's surviving offspring. Reproductive success depends upon all those factors that contribute to the *survival and mating success of an individual. These include *mate selection, courtship success, fertility, and fecundity. In studying fitness in the natural environment it is usually necessary to use an index of reproductive success, but such indices only approximate to true fitness, because they take no account of the viability of the offspring. In particular, in investigating the *function of a trait, the increase in reproductive success that the trait confers on its possessor's *genetic constitution is what needs to be established. For example, the function of *nest-building is to provide a nest to keep the occupants warm, and to help protect them from predation. This implies that these consequences of nest-building confer higher reproductive success on the genotype of the individuals that practise it effectively.

reproductive value A age-related indicator of *reproductive success. The *fitness of a genotype in a Darwinian sense can be measured in terms of the number of its progeny. The age-specific reproductive value is an index of the extent to which the members of a given age group contribute to the next generation, between now and when they die.

resilience The extent to which an activity is resistant to pressures of time. Low-resilience activities are those that are curtailed when time is short, because other (high-resilience) activities take priority. An activity that is abandoned under these conditions is often regarded as a *leisure activity.

The concept of behavioural resilience is closely related to elasticity of *demand. An indication of the relative importance of the various activities in the animal's repertoire is given by the degree to which the activity is performed when there is insufficient time or energy to carry out all those activities that would normally occur in a day.

response An overt (usually) behavioural reaction to a stimulus. However, release of *pheromones, pupil enlargement, and other covert reactions, are sometimes counted as responses. Strictly, the activation of any effector could be counted as a response.

reward Synonymous with positive *reinforcement.

rhythms Cyclic aspects of behaviour. The period of a behavioural rhythm may be as long as a year, or as short as a few minutes. *circannual rhythms are endogenous and usually have (in isolation) a periodicity slightly less than 365 days. The endogenous rhythm is due to an internal *biological clock. *circadian rhythms usually have (in isolation) slightly less than a 24-hour periodicity. These are also driven by an internal biological clock.

Tidal rhythms, found in many marine organisms, have a period related to the lunar cycle of 29.5 days. The tidal cycle is repeated twice in the lunar cycle. Thus for about 7 days after the new or full moon the tides diminish in amplitude, and then increase for a further 7 days. This general pattern is the same around the world, but can be modified by local conditions. Many marine animals follow their local tidal cycle, even when isolated in the laboratory.

Short-term rhythms of feeding, drinking, preening, etc. are influenced by the endogenous biological clock, but also have a strong exogenous component. Thus while the *feeding rhythm is influenced by the waxing and waning of *hunger, mathematical analysis of meal patterns often shows that some independent timing factor is also involved. Physiological **biorhythms** influence many types of behaviour, including *sleep, *defence of *territory (in birds), and *predatory behaviour attuned to the daily routines of prey.

ritualization A process of *evolution by which behaviour patterns become modified to serve a *communication function. When it is of advantage to an animal that some aspect of its behaviour provides information to another animal, then *natural selection operates to transform the behaviour pattern into a more reliable and conspicuous form of communication. Usually the behaviour pattern becomes stereotyped in form and incomplete in its execution. For example, ritualized grooming is often restricted to a certain part of the body, and the movements themselves become curtailed, and are sometimes merely token movements.

Ritualized behaviour often has a **typical intensity**. For example, the black woodpecker (*Dryocopus martius*) drums against dry branches to indicate to other birds that its *territory is occupied, and to attract females. This drumming has a characteristic rhythm, that is stereotyped compared with the sound made by chipping out a nest cavity, from which it is clearly derived. The typical intensity of ritualized drumming makes it unambiguously different from the normal drumming.

Ritualized behaviour involves changes in *motivation. A well-known example is *courtship feeding in birds. Females often beg food from their partners during courtship, with behaviour that is otherwise seen only in food begging in juveniles. The female is not especially hungry at the time, the behaviour is clearly ritualized, and has different motivation from normal food begging.

The evolution of ritualized behaviour can be traced to a variety of types of everyday activity. These usually have the common feature that they are already potential sources of information to other animals. Such activities include *conflict behaviour, *intention movements, and *displacement activities.

S

scent marking Behaviour that aids the dispersal of *pheromones. Many animals, including humans, have specialized glands that produce scented secretions, or pheromones. These may be released in a volatile form into the air, or they my be actively deposited by the animal on to the ground, or a suitable object, by means of behaviour known as scent marking.

Pheromones are a means of chemical *communication, and the advantage of depositing the scent, rather than releasing it into the air, is that the scent disperses at a lower rate, and will persist for hours in the absence of the animal that produced it. Scent marking thus provides a relatively permanent signal, especially effective in certain situations, such as the marking of paths or *territory boundaries.

It is probable that all mammals have scent glands, and they are also found in fish, reptiles, and amphibians, though not, it seems, in birds. Scent marking occurs in many insects, including ants (Hymenoptera), bees (Apidae), and wasps (Vespoidea). In the former, it is used in trail marking, and in the latter for marking the entrance to the nest.

In mammals, the scent glands are located on various parts of the body, depending upon the species. Thus in deer (Cervidae), and other ruminants, the glands occur between the toes and on the forehead. In bats (Chiroptera), they may be on the throat, lips, and wings. Many species have scent glands in the genital region, and some use faeces and urine as scent carriers.

Scent marking is an important aspect of *social organization in many animals, and is involved in a wide range of activities, including the coordination of *foraging in ants, *alarm signals in fish, and the *sexual behaviour of many mammals.

schooling A form of *social organization in fish. Like *herding in mammals and *flocking in birds, schooling confers advantages primarily in connection with predators. Most schooling fish are relatively small, and few predatory fish school, with the notable exception of the Pacific barracuda (*Sphyraena argentea*). Tuna (*Euthynus pelamis*) school when small, but the large animals are solitary.

A predator is less likely to encounter a fish of a prey species if the prey stay clumped rather than disperse randomly. Fish in schools are harder to target, due to the **confusion effect**. Movements of other fish visually distract a predator trying to keep track of one particular prey. Moreover, the smaller fish in a school can turn more quickly than the larger predator, although they cannot swim as fast. They are thus able to execute 'rehearsed' avoidance manoeuvres.

In addition to anti-predator functions, schools may have a number of other functions. There are hydrodynamic advantages of swimming in a group, possible navigation advantages during *migration, and improved *vigilance in respect of feeding opportunities and obstacle avoidance.

search image A supposed mental image (of a sought-after object) that an animal may have while *searching. The evidence for this is circumstantial. Field studies can reveal the extent to which *foraging animals are being selective, and overlooking some perfectly edible items. Observations in the field seem to show that predatory birds concentrate on one type of prey within a given foraging period. In the laboratory, studies of domestic chicks (*Gallus gallus domesticus*) have shown that they demonstrate a definite improvement with experience in their ability to detect camouflaged grains of rice. Carrion crows (*Corvus corone*) have been shown to respond selectively to a particular type of prey, even when other equally likely types are available. The birds were trained to search among beach stones for pieces of meat that had been hidden under mussel shells painted various colours. The birds would concentrate on shells of one colour, ignoring others. If a bird turned over a shell of a particular colour and found no meat, then it would usually start searching for shells of another colour. Similar phenomena have been demonstrated in laboratory experiments on a number of species of fish, bird, and mammal. However, none of these studies show that the animals have a mental image as such, and the results can also be explained in terms of the much-studied topic of selective *attention.

searching An aspect of *foraging (although also sometimes a precursor of *courtship). Usually a searching animal will confine its activities to a particular locality or *habitat. Many animals learn to concentrate their search in areas in which they have previously been successful. Having arrived in a particular locality, there are various searching strategies that can be employed. Random search is generally the least efficient of these, partly because it often results in the animal revisiting areas that have already been covered. The best searching strategy is attuned to the distribution of prey. If the prey is widely scattered, then it is better to make fewer changes of direction and cover more distance per unit time. If the prey is clumped into patches of high density, it is better to make more changes of direction in the vicinity of a capture. Blackbirds (*Turdus merula*), searching for prey in open meadowland, tend to concentrate on the surrounding area just after a capture, and this tendency is more marked if the prey is patchily distributed. Thus the blackbird is able to alter its search strategy according to the circumstances.

Once an area has been thoroughly searched, it is better to move to another area, but the period before a fruitful area should be revisited depends upon the rate of prey replenishment. Thus wagtails (*Motacilla* sp.) foraging on insects washed up on a river bank systematically exploit a particular stretch of river bank that they defend as a feeding *territory.

selective breeding The practice of selecting animals for breeding on the basis of phenotypic characteristics such as their appearance, performance, or production records. This is common in breeding dogs (Caninae) for showing, horses (Equidae) for racing, and farm animals for milk or meat production. Selective breeding may produce dramatic changes in the characteristics of animals, but it cannot produce entirely new ones. What is changed is the proportion of relatively desirable traits. Selective breeding has been important in the domestication of animals.

self-awareness A mental state in which the individual is aware of itself as having knowledge about itself. A dog may be aware that it is known by its name 'Rover'. It may respond to the name Rover in such a way that indicates that it distinguishes between itself and other dogs (Caninae). Thus it may sit on the command 'Rover sit', but not to the command 'Fido sit', even when it is in the presence of another dog with the name 'Fido'. However, such responses do not show that the dog *knows that* its name is Rover. They do not show that the dog had explicit *knowledge of itself. If, however, the Rover looked in a mirror and knew that the image in the mirror was that of the dog called Rover, then it would have explicit knowledge about itself. It would have self-awareness.

Humans (above a certain age), chimpanzees (*Pan* sp.) and orang-utans (*Pongo* sp.), and possibly gorillas (*Gorilla* sp.) are claimed by some to have passed such mirror tests (e.g. they have investigated marks on parts of the body that they can see only through a mirror). Other primates, and other animals, have (so far) not passed the mirror tests.

An animal that possessed self-awareness might be expected to have some insight into the mental lives of other members of its own species. This line of thought gives rise to the topic, called *theories of mind, and to experiments that aim to demonstrate that some animals do have such abilities.

sense organs Organs adapted for the reception of stimuli or events outside the nervous system. The stimuli may be external to the animal as a whole, such as visual or auditory stimuli, or they may be internal to the body, such as pressure within a tissue, or change in the length of a muscle. Each sense organ contains some kind of transducer, that converts light, chemical, and mechanical energy, etc., into the type of electrical energy that can be registered by the nervous system. This registration is done by *sensory receptors, specialized nerve cells that are part of the sense organ and that connect to other cells in the nervous system.

sensitive periods Periods of time during which a developing animal is especially sensitive to particular types of experience. For example, in the development of *song in certain birds, there is a period during which the young bird is capable of *learning the song of its parents, or foster parents, even though it does not sing itself for many months.

Sensitive periods are especially important in *imprinting, the process by which social preferences are influenced by experience. The sensitive period usually begins at birth, or hatching, and the juvenile's social preferences become progressively narrowed to the point where they are no longer influenced by experience. In the case of filial imprinting, the young animal is predisposed to become attached to objects that provide a particular range of stimuli, and in the natural environment these are usually provided by the parents or siblings. In the laboratory, the young animal can be induced to develop an attachment to unnatural objects, such as a cardboard box, rubber ball, or a member of another species. The sensitive period for filial imprinting usually terminates before that for sexual imprinting. Thus ducks (Anatidae) of one species that are reared by ducks of another species will subsequently attempt *courtship with members of the foster species. Sensitive periods are also important in the development of *habitat and *food preferences.

sensitization Increase in the probability of a *response resulting from repeated presentation of a biologically significant stimulus. Whereas *habituation decreases the response probability, and occurs to relatively insignificant stimuli, such as changes in illumination, sensitization results from repeated significant stimulation, such as electric shock or presentation of food. For example, the probability of an octopus (*Octopus* sp.) emerging from its home to attack a neutral stimulus, such as a white disc, increases if it has recently been given food. The results of sensitization can easily be confused with those of *conditioning, or other forms of *learning. However, sensitization occurs in the absence of correlated *reinforcement, and is a form of learning more akin to habituation.

sensory receptors Specialized nerve cells that are responsible for the transduction and transmission of information within *sense organs and other sensory structures. Sensory receptors are specialized according to the environmental energy to which they react. Thus mechanoreceptors undergo electrochemical changes in response to deformation of the cell membrane. It is characteristic of sensory receptors that the environmental energy is converted into a graded electrical potential, called the generator potential. This is usually proportional to the intensity of stimulation of the receptor. When the generator potential reaches a certain threshold level, it triggers an action potential that travels along the axon of the nerve cell. This is the transmission part of the sensory process, and the information is usually coded such that the more intense the stimulus, the faster the rise in generator potential, and the higher the frequency of action potentials that travel along the neuronal axon.

In the absence of stimulation, the generator potential falls gradually to its resting level. When it falls below the threshold level, no further action potentials are generated. When stimulation is resumed, there may be a short delay, called the refractory period, while the generator potential rises from its resting level to the threshold level. Overall, the intensity of stimulation is coded in terms of the frequency of action potentials in the sensory receptor axon.

sensory systems Systems involving sensory receptors, sense organs, sensory (afferent) nerves, and sensory receptive fields in the *brain, that enable sensory *perception. The most common sensory modalities are *chemoreception (*taste and *smell), *hearing, and *vision, but electromagnetic *sensitivity, *mechanoreception, *thermoreception, and *time are also important in many animals.

sexual behaviour All behaviour leading to the fertilization of eggs by sperm. Once fertilization has taken place, there may be further sexual behaviour directed towards further fertilizations, or there may be a switch to *parental behaviour. Sexual behaviour may involve *copulation, leading to internal or external fertilization, or copulation may be absent. In the three-spined stickleback (*Gasterosteus aculeatus*), for example, the female swims into the nest, deposits her eggs, and swims out. The male follows the female and fertilizes the eggs.

Every sexually reproducing species must meet a number of criteria. These include mate selection, reproductive timing, coordination during mating, and promotion of viable offspring. Although species vary greatly, each has solved the same basic problems during the course of *evolution.

*Mate selection is of prime importance in preventing interbreeding between species, and most species have characteristic *isolating mechanisms. Different species have different *mating systems, but mate selection is always subject to the same evolutionary pressure. The sexual partner should provide the best possible benefit for the offspring, both in terms of genetic contribution and in terms of *parental care.

Reproductive timing is largely determined by the *habitat, because conditions must be favourable for the extra food requirements of the growing young. Thus, seasonal factors are the main determinants of the *sexual cycle that is characteristic of each species. Coordination and *cooperation between the sexes is also important when the sexes live apart for part of the season. Finally, promotion of viable offspring is largely a matter of parental care, which is especially well developed in *incubation in birds and *maternal behaviour in mammals.

sexual cycle A cycle of sexual activity that usually has a period of a year. It is largely under influence of a *biological clock that controls the release of *hormones, but exogenous factors to do with the weather may also play a part. Many animals have an annual breeding season, within which there may be a shorter **oestrous cycle** of female receptivity. Ovulation occurs during the cycle and the female becomes receptive to mating. Fertilization can take place only during this period. The oestrous cycle of the female rat (*Rattus norvegicus*) lasts 4–5 days, and may be repeated if fertilization does not take place. Domestic dogs (Caninae) have two periods of heat (oestrous cycle) per year, one in spring and one in autumn. In some species, *copulation is necessary for ovulation. These include the rabbit (*Oryctolagus cuniculus*), cats (*Felis catus*), and the short-tailed shrew (*Blarina brevicauda*). Some primates, including humans, have a different system, called the *menstrual cycle.

sexual selection A form of *natural selection, which depends upon the advantage that certain individuals have over others of the same sex and species solely in respect of reproduction. There are two ways in which a male can gain an advantage over other males. First, they can compete directly with one another by *fighting, or by some type of ritualized combat. This is sometimes called **intrasexual selection** (selection within a sex) or **male rivalry**. Secondly, males can compete indirectly in attracting females by means of special displays and adornments, sometimes called **intersexual selection** (selection between sexes) or **female choice**.

Male rivalry can be exemplified by the red deer (*Cervus elaphus*). The stags grow antlers each year and, in the autumn rutting season, they directly challenge each other for ownership of females, which have no antlers and which are herded into harems by the successful males. The females have little or no choice of sexual partner, because the males defend their harems against rival males.

Female choice can be exemplified by the peacock (*Pavo cristatus*). The males develop enormous ornamental tails that they *display during *courtship. Females prefer males with good tails, even though they are costly to produce and maintain, and are likely to hinder escape from predators. There are various theories as to how this female preference is maintained in the face of these pressures of natural selection. One theory is that females prefer large tails because they indicate that the male must be fit (and free from parasites) to carry such a handicap. Another is that females that mate with attractive males are likely to have attractive sons, who will also have high reproductive success. The detailed evolutionary theory advanced to account for sexual selection remains controversial.

shaping Training an animal to perform a certain act by progressively manipulating its behaviour by providing *reinforcement of activity in the direction of the act. For example, to train a rat (*Rattus norvegicus*) to press a bar, the animal is first rewarded for approaching the bar, then for touching the bar with a forelimb, then for pressing the bar, then only for pressing the bar sufficiently hard to operate an automatic reward mechanism.

sign stimuli A part of a stimulus configuration that is external to the animal and relevant to a particular response. For example, the male three-spined stickleback (*Gasterosteus aculeatus*) has a characteristic red belly when in breeding condition. This is a sign stimulus that elicits aggression in other territorial males. A crude wooden model, appropriately coloured with a red underside, is more effective than a real stickleback without a red belly. Thus many of the details and structure of a male rival are apparently ignored by other males. The red coloration is much more effective if it is on the underside of the model fish.

Responses to sign stimuli may depend partly upon the animal's *motivation. Thus the sign stimuli by which a herring gull (*Larus argentatus*) identifies an egg (i.e. size, shape, coloration, etc.) that has rolled out of the nest may elicit a retrieval response in one motivational context, while the gull may eat the egg in another context.

sleep A type of *dormancy, found in birds and mammals, during which there is a characteristic type of electrical brain activity. Measurements of the gross electrical activity of the brain produce an electroencephalograph (EEG), which shows a characteristic **alpha rhythm** of about 8–12 Hz (in humans). If the animal awakes, the rhythm becomes desynchronized. In deep sleep there are low-frequency (3–4 Hz), high-amplitude **delta rhythms**. During deep sleep there may also be periods of low-amplitude fast EEG activity. This appears to be paradoxical, because the pattern is typical of the awake animal. If humans are woken during this type of sleep, they report dreams. No dreams are reported when the subject is awakened at other times. During such sleep paradoxical rapid eye movements (REM) occur, and the animal may vocalize or make incipient locomotary movements. Modern nomenclature distinguishes between **quiet sleep** (QS), characterized by rhythmic EEG, and **active sleep** (AS), characterized by a desynchronized EEG and signs of dreaming. The **sleep cycle** (QS+AS) tends to be a regular rhythm that is characteristic of the species. There is a strong correlation between the period of the sleep cycle and adult body weight and basal metabolic rate.

Mammals and birds usually seek out a typical sleep site and adopt a typical sleep posture. However, there is considerable variation among species. Total sleep time (in 24 h) ranges from 20 h in the opossum (Didelphidae) to 2 h in the giraffe (Giraffidae). The duration of sleep episodes ranges from 20 min in sheep (*Ovis*) to a number of hours in carnivores. The human average is about 8 h, but some people habitually sleep 16 h/day and others only 2 h. A daily sleep ration of 45 min has been reliably recorded in an otherwise normal person.

The function(s) of sleep are controversial. It is difficult to argue that sleep serves a vital restorative function when some animals and some people sleep so little. It is difficult to argue that sleep is merely an energy-saving time-filler, when sleep deprivation leads to a certain amount of catching up. To add to the mystery, it is evident that there is some link between sleep and *thermoregulation.

smell *See* Olfaction.

social dominance *See* Dominance.

social facilitation A form of apparent *imitation, that may occur as a result of a tendency to investigate places where other members of the species have been observed. For example, if a lone chicken (*Gallus gallus domesticus*) is allowed to eat until completely satiated, and is then introduced to others that are still feeding, the satiated bird will resume eating. Many animals have been shown to eat more when fed in groups than when fed alone. This has been demonstrated experimentally in fish, chickens, opossums (Didelphidae), and domestic puppies (Caninae).

A tendency to concentrate on the type of food being eaten by others has been observed in sparrows (Passerinae) and chaffinches (*Fringilla coelebs*). *Avoidance of noxious food can also be socially facilitated. Attempts to kill large flocks of crows (*Corvus*) in the USA, by providing poisoned bait, were not successful because the majority avoided the bait after a few individuals had been poisoned. Similar *bait shyness has been documented in rats (*Rattus norvegicus*).

social interactions The result of an animal reacting to stimuli provided by another member of the same species. In some cases these stimuli may be relatively simple *sign stimuli, as in the defence of *territory by sticklebacks (*Gasterosteus aculeatus*) (the red breast of the rival male is the sign stimulus) and the feeding of nestlings by songbirds (distinctive markings within the gape provide the sign stimuli). In other cases the stimuli may be more subtle and complex, requiring some *learning or *cognition on the part of the participants, as in some cases of *dominance and *deceit.

social learning *learning in a social context. This includes learning about the *dominance status, *motivation, *goals, and abilities of other members of the group. It includes learning in situations where the *reinforcement is social, such as the receipt of food items, *grooming, or proximity to a superior. Social learning manifests itself in *emulation and *imitation of others, and sometimes in apparent *deceit.

social organization The social structure that results from *social interactions and *social relationships. *Dominance and family relationships form the basis of social organization in many animals, but in some species (e.g. ants (Hymenoptera) and termites (Isoptera)) the social structure may be due to a genetically based caste system. In some colonial invertebrates, such as the corals (*Corallium*) and bryozoans (Bryozoa), the division of labour amongst the semi-independent zooids

(Ascidiacea) may be so marked that the colony is regarded by some as a single organism, akin to an organized society.

To some extent, all animals have some degree of social organization, even if only that necessary for *sexual behaviour. Some species show social organization in some aspects of behaviour, but not others. Examples can be found in the *schooling behaviour of fishes, and the *alarm responses of many birds.

social relationship A relationship between two individuals who are well known to each other, and who mould some of their behaviour in accordance with each other's individual characteristics. A social relationship found in many species is that between parent and offspring. Examples can be seen in the *maternal behaviour of mammals, and the *parental behaviour of some birds. In other species, there may be no social relationship of this type. In very many species, however, some kind of social relationship is developed during *courtship and *sexual behaviour. Other types of social behaviour that may involve the development of relationships include social *dominance and *cooperation.

sollwert A representation of a *goal-to-be-achieved in a negative *feedback system.

song An type of *communication in birds (a term sometimes applied to other animals, such as whales (Cetacea)). Among birds, there is a wide range of types of vocal learning. In some, such as cuckoos (*Cuculus* sp.) and doves (*Streptopelia* sp.), the development of vocalizations is not influenced by early experience. It is said to be *innate. Others, such as the white-crowned sparrow (*Zonotrichia leucophrys*), can learn modifications of the song of their own species, but cannot learn parts of the song of other species. Other birds, such as meadowlarks (*Sturnella* sp.), do not normally imitate songs of other birds in nature, but are capable of it under laboratory conditions. Others, including mocking-birds (Mimidae), lyre-birds (Menuridae), some starlings (Sturnidae), and corvids (Corvidae), exhibit true *imitation of the songs of other species under natural conditions.

The prime function of bird song is the defence of *territory. In some species it also serves as a means by which mated pairs keep contact in conditions of poor visibility. Bird song is a specialized type of *vocalization, evolved to serve specific functions, and is distinct from other types of bird vocalization such as *alarm calls.

specific hunger A type of *hunger that is satisfied by specific dietary requirements, such as vitamins and minerals. Many animals vary their food intake according to the nutritive value of the products of digestion. A variety of mechanisms are involved in this type of regulation. The simplest mechanism is the direct detection of the substance in the food, as is the case with sodium. Animals can detect sodium in the diet in two main ways. First, sodium salt (NaCl) is a primary aspect of *taste in most vertebrates. Secondly, sodium has profound effects upon the body fluids, and its presence there can be directly detected. Sodium appetite appears to be *innate, but many animals are adept at *learning and remembering the location of sources of sodium.

There are many vitamins and minerals that animals are not able to detect, either by taste or by their levels in the blood. Nevertheless, deficient animals develop strong preferences for foods containing the missing substances. Rats (*Rattus norvegicus*) deficient in thiamine show an immediate marked preference for a novel food, even when that food is thiamine deficient. The preference is short lived. If consumption of a novel food is followed by recovery from the dietary deficiency, however, then the rat rapidly learns to prefer the novel food. Such rapid learning on the basis of the physiological consequences of ingestion enables the rat to exploit new sources of food, and to find out which contains the required ingredients.

The effects of a vitamin-deficient diet have much in common with *poison avoidance. Vitamin-deficient rats are reluctant to eat familiar food, and show a more than normal interest in novel foods. The aversion to previously familiar food persists even after the animals have recovered from the deficiency. Rats that become sick after eating poisoned food also show an aversion to familiar food and an interest in novel foods.

sperm competition Competition for fertilization, as distinct from competition for access to females. In many species, females mate with more than one male within a short space of time. Sperm from both, or either, may be successful in fertilizing her eggs. In dungflies (*Scatophaga stercoraria*), sperm from the second male displace most of those from the first. In mammals, when two males copulate with the same female, the male with the larger number of sperm usually ensures paternity of a larger proportion of the female's litter. Among primates, sperm competition is likely to be important in species where females are routinely mated by more than one male. Such species have larger testes for their body size than do harem-holding or monogamous species.

Precautions against sperm competition include copulation in private, away from possible interference, as is seen in some mammals. Amongst insects, it is common for the male to remain in the copulatory position for a number of hours, other forms of mate guarding also occur in the dungfly and other species.

startle response A *reflex response to sudden stimulation. In vertebrates, it is usually accompanied by *autonomic arousal, and followed by an *orienting response.

stereotyped behaviour Any behaviour that is repetitive and relatively fixed in form. Examples include *stereotypies, habitual or routine behaviour, and behaviour that is necessarily repetitive, such as digging. Stereotyped behaviour is not necessarily *abnormal, but it often occurs in captive animals.

stereotypies A type of repetitive behaviour that appears under conditions of *stress. An example is repetitive sham chewing by tethered sows (*Sus*). Stereotypies are an important aspect of the study of animal *welfare.

stigmergy The production of behaviour that is a direct consequence of the effects produced in the local environment by previous behaviour. For example, when termites (Isoptera) start to build a nest, they modify their local environment by making little mud balls, each of which is impregnated by a *pheromone. Initially the termites deposit their mud balls at random. The probability of depositing one on top of another increases as the sensed concentration of pheromone increases. After the first few random placements, the other termites tend to deposit their mud balls in the same place, so that small columns are formed. The pheromone from neighbouring columns causes the tops of the mud columns to lean towards neighbouring columns, Eventually the tops meet, forming arches, the basic building units of the nest. As other stigmergic rules come into play, involving water vapour, carbon dioxide concentration, and the presence of the queen, the whole complex nest structure is produced. This may include the royal cell, brood nurseries, air conditioning, larders, and communication tunnels.

The essence of stigmergy is that simple behavioural rules change the environment in such a way that new rules are triggered by the new environmental stimuli. Thus the behavioural repertoire becomes self-organizing. This is the principle behind much of the social behaviour of ants (Hymenoptera) and termites.

stimulus filtering A series of processes by which the numerous stimuli that impinge on an animal at any one time are prevented from influencing its behaviour. Much more information is potentially available to an animal than it could possibly register and respond to. Some stimulus filtering is inherent in the limited capabilities of the *sense organs, but other stimuli that the animal is capable of detecting are ignored, as is typically the case with *sign stimuli. The animal's *motivation may then determine which aspects of the stimulus situation receive *attention. Finally, the stimuli to which the animal responds may depend upon the outcome of some *decision-making process.

stimulus relevance A principle that asserts that the strength of an *association between a stimulus and some consequences depends partly upon the nature of the consequences. Thus a rat (*Rattus norvegicus*) readily associates a light or sound with electric shock consequences, but does not associate taste stimuli with such consequences. On the other hand, the rat strongly associated taste stimuli with subsequent poisoning, but does not associate light or sound stimuli with such consequences.

stimulus–response (S–R) and stimulus–stimulus (S–S) theories Theories of *learning that claim that what is learned is a connection between stimulus and response (the S–R type of theory), or a connection between two stimuli (the S–S type of theory). The latter amounts to the claim that the animal forms an *association between two stimuli.

stress An aspect of *motivation that results from specific internal or external stimuli, called **stressors**. These can really only be identified by their effects upon the animal's physiology. It appears that invertebrates have specific automatic responses to single natural stressors, and lack a general mechanism that could be implicated in the response to a variety of stressors. Stress in vertebrates is closely associated with *hormones, levels of which are the main physiological measure used to indicate stress. Most behavioural measures of stress involve some kind of *avoidance, indicating that the condition is aversive. Failure to cope with stress causes *distress.

Common stressors include the presence, or signs of, predators; exposure to *novelty; mother–infant separation, conspecific *fighting, and social subordination (*see* DOMINANCE). Environmental factors, such as cold, can also be stressors. The syndrome of physiological changes induced by stressors is known as the **general adaptation syndrome**. This syndrome consists of specific physiological changes, that result from a diverse range of stimuli (the stressors) and have important effects upon physiological equilibria and *homeostasis.

suffering An aspect of *motivation that is generally considered to be aversive, and may cause *stress. Conceptually, suffering has much in common with *pain. The problem of assessing pain or suffering in animals is that it can be done only by analogy with ourselves, and this approach is open to the objection that it involves *anthropomorphism.

Do animals have to be conscious to suffer? At the commonsense level, we are inclined to suppose that they do. When we are unconscious we do not suffer pain, or mental anguish, presumably because parts of the brain are deactivated. However, we have no conception of what *consciousness in animals might involve, if it exists. Therefore we can draw no conclusions about the relationship between consciousness and suffering (or pain) in animals, except by analogy with ourselves. Some animals, especially mammals, have similar behavioural and physiological reactions to stimuli that would induce pain or suffering in us, and we would normally assume that the presence of such reactions indicates that the animal is in pain. The problem is confounded by the fact that non-mammalian animals, with nervous systems very different from ours, show characteristic behavioural reactions to pain stimuli, but their physiological reactions are somewhat different from ours. In the interests of animal *welfare, we can give animals the benefit of the doubt, but we have no way of finding out whether their inner experiences of suffering are similar to ours, or whether they suffer at all. Moreover, the concept of suffering is not really necessary to explain animals' *avoidance reactions to *stress, or to situations likely to cause pain.

sunbathing An aspect of *thermoregulation in which the animal controls its body temperature by selective exposure to the sun. It is common in land invertebrates and cold-blooded vertebrates. For example, the desert locust (*Schistocerca gregaria*) maintains a perpendicular *orientation to the sun during the morning and evening, when the air is cold, but turns parallel to the sun's rays during the day when the air is hot. Similarly, after cold nights, lizards (*Draco* spp.) often seek sunlight, basking in the sun so as to raise their body temperature.

Sunbathing in birds, such as pigeons (*Columbia livia*), is a response to light rather than to the heat of the sun. They have a typical sunbathing posture, in which they spread out their wings and ruffle their feathers. This is quite a different posture from that involved in thermoregulation. Sunbathing in birds is designed to allow the sunlight to reach the skin, where it encourages the synthesis of vitamin D. Many mammals show a preference for resting in sunlight, provided that it is not too hot. The solar radiation provides heat that would otherwise have to be provided by metabolic processes.

sun navigation The use of information about the position and movement of the sun for *navigation. Provided with information about the *time in location x, and observing the sun in location y, and animal would have enough information to work out the relative positions (latitude and longitude) of locations x and y. In the northern hemisphere, if the sun is later in the day at location y than it would be at location x at a particular (known) time, then y must be west of x, and vice versa. If the sun is higher in the sky at noon at location y than it would be at x, then y is south of x, and vice versa. In theory, the sun (provided the time at location x is known) provides sufficient information for bicoordinate navigation, but there is no conclusive evidence of these abilities in animals. There is evidence that navigating birds use the sun and stars as a compass, but a compass alone is not sufficient for bicoordinate navigation. Similarly, there is evidence for use of a magnetic compass by animals, but this alone is not sufficient for bicoordinate (i.e. map and compass) navigation. Despite much research, there is no convincing evidence for the use of a map analogue by animals.

Solar information gained on an outward journey can be used to navigate during the return journey. For example, *Cataglyphis* ants of the Sahara desert navigate by path integration. While foraging away from home, they continually measure all distances covered and all angles turned, and integrate these linear and angular components into a continually updated vector that points towards home. *Cataglyphis* uses a compass to monitor the angular components of its movements. This compass utilizes the pattern of polarized sunlight in the sky, as do some bees. The photoreceptors in a particular region of the ant's (and bee's (Apidae)) eye are sensitive to directional oscillation of this pattern. The pattern of polarization changes with the elevation of the sun above the horizon. The ant is equipped with a neural template that resembles the polarization pattern when the sun is at the horizon, but differs from it for all other elevations. The best possible match between the template and the external pattern is achieved when the insect is aligned with the solar (or anti-solar) meridian. As the animal rotates about its vertical body axis, the match decreases systematically. The best match gives the zero point of the compass, and this deteriorates as the ant selects other compass directions.

supernormal stimulus A stimulus that surpasses a natural stimulus in its effectiveness. Thus herring gulls (*Larus argentatus*) prefer to retrieve larger than normal (dummy) eggs into their nest. Animals often have an open-ended preference for stimulus properties, such as size, sweetness, or colour, that is limited in the natural world. When presented with an artificial supernormal stimulus, they prefer it to the natural one.

survival value The survival of a trait within a population depends upon the extent to which the trait contributes to *reproductive success, which depends partly upon the selective pressures inherent in the environment. A number of features of the environment could jeopardize reproductive success, by leading to the death of the parent by starvation, predation, failure to breed as a result of *competition for mates or nesting sites, or failure of the young to survive due to lack of *parental care, food, or protection from predators.

We can think of the animal as designed by *natural selection to fulfil certain *functions, or roles of traits in the survival and reproductive success of the individual animal. The survival value of a trait is thus a measure of the importance of its role in the animal's life.

symbiosis The living together of organisms of different species to their mutual benefit. It occurs amongst plants, between plants and animals, and among animal species. It can occur as a relationship between individuals, between individuals and societies, and even between societies.

Many aphids (Aphidae), and related species, gain protection by associating with ants, while the ants benefit by obtaining food. Thus when the garden ant (*Lasius niger*) encounters a bean aphid (*Aphis fabae*) it caresses the aphid with its antennae. This induces the aphid to exude from its anus a by-product of digestion, called honeydew, which the ant consumes.

The honey badger (*Mellivora capensis*) lives in symbiosis with a small bird called the black-throated honey guide (*Indicator indicator*). When the bird discovers a hive of wild bees, it searches for a badger and guides it to the hive by means of a special *display. The badger opens the hive with its large claws, protected from the bees (Apidae) by its thick skin. It then feeds upon the honeycombs, while the bird gains access to the bee larvae and wax, upon which it feeds. If the honey guide cannot find a badger, it transfers its attention to the next best alternative, humans.

T

taste A form of *chemoreception in which there is direct contact between the chemoreceptors and relatively concentrated solutions of chemical substances. For example, chemoreceptors in the feet of blowflies (Calliphoridae) can detect salt, sugar, and pure water. In vertebrates, the sense of taste is relayed via the facial (VII) and glossopharyngeal (IX) cranial nerves, while the sense of smell is transmitted by the olfactory nerve (I). The basic tastes of mammals are acid, bitter, salt, and sweet. In humans different parts of the tongue are sensitive to different tastes. The flavour of food depends upon both taste and smell (*see* OLFACTION).

taxis A type of *orientation, based upon directional information, in which the animal heads directly towards or away from the source of stimulation. This may be achieved by means of simple *sensory receptors and successive comparison of stimulus intensity in different locations, achieved by turning movements. For example, when the larva of the house fly (*Musca domestica*) has finished feeding, it seeks out a dark place where it pupates. At this stage it will crawl directly away from a light source, and it is said to show **negative phototaxis**. The maggot has primitive eyes on its head that are capable of registering changes in light intensity but that are not able to provide information about the direction of the light source. As the maggot crawls, it moves its head from side to side. When the light on the left is brighter than on the right, the maggot is less likely to turn its head towards the left. It thus tends to crawl more towards the right, away from the light source. In response to an increased level of illumination, the maggot increases its rate of head turning. Thus, a form of orientation is achieved by successive comparison of the stimulus intensity in different nearby places. The rate of turning, or sampling the environment, varies with the stimulus intensity, and this is known as *klinotaxis. This type of taxis is common in response to gradients of chemical stimulation.

Simultaneous comparison of the intensity of stimulation received at two or more receptors enables the animal to strike a balance between them. It can then achieve *tropotaxis, which enables it to steer a course directly towards, or away from, the source of stimulation. Eyes that are capable of providing information about the direction of light by virtue of their structure are capable of *telotaxis, a form of directional orientation that does not depend upon simultaneous comparison of the stimulation from two receptors. *Menotaxis is a form of telotaxis that involves orientation at an angle to the direction of stimulation.

teaching A form of *parental behaviour (usually), in which the young learn from the behaviour of their elders. In some species, the young learn from the behaviour of their elders by *imitation, *social facilitation, and *tradition. None of these involve teaching, because the animal whose behaviour is copied takes no account of whether it is watched or not. Teaching is involved when an animal modifies its behaviour, with no immediate benefit to itself, only in the presence of a naïve observer, with the result that the observer gains knowledge, or learns a skill with greater efficiency than otherwise.

Whether or not such teaching occurs in animals, other than humans, is a matter of controversy. True teaching has been claimed for carnivores, primates, and raptors. Some carnivores and raptors appear to teach their offspring to kill, by releasing live prey in their presence. Teaching of *tool use has been claimed for primates. Teaching implies some *theory of mind, not a universally accepted theory, and for this reason it is controversial.

telotaxis A type of directional *orientation that does not depend upon simultaneous comparison of the stimulation from two receptors, but which requires eyes (or ears) that are capable of providing directional information by virtue of their structure. When there are two sources of stimulation, the animal moves towards one, and never in a median direction. This shows that the influence of some of the stimuli is inhibited.

tendency The level of underlying motivational potential to perform an activity. An animal may have simultaneous tendencies to perform incompatible activities (such as fighting and fleeing), but it can perform only one at a time. Which activity is performed is a question of *decision-making. When two tendencies are nearly equal in strength, *conflict behaviour may occur, but usually the *motivation to perform one activity is sufficiently greater than the alternatives, for that activity to take precedence over the others. The strength of tendency to perform a particular activity is often a combination of internal and external factors. For example, feeding tendency is often a combination of *hunger and the perceived availability of food, its apparent palatability, etc.

territorial behaviour A form of *defence of food sources, nest-sites, or other resources, against other members of the same species. Many animals defend patches of ground against intruders, usually of the same species. This is sometimes accomplished by outright *aggression towards intruders, sometimes by *threat displays, by *scent marking (in mammals) and by *song (in birds).

Territorial defence has both costs and benefits, and animals defend territories only when it is economical to do so. For example, spotted hyenas (*Crocuta crocuta*) defend territories in the Ngorongoro crater, where their prey is predictable and abundant, but not in the Serengeti plains, where it is very seasonal. The hyenas in the Serengeti wander over a wide *home range, but do not defend any territory. Many animals vary their defence strategies depending upon the food supply. When food levels become high intruder pressure increases, and both sunbirds (Nectariniidae) and squirrels (Sciuridae) give up their territories, because the defence costs become too great. At the other extreme, *foraging productivity may become so low that, even with territorial defence, the animal is unable to meet its daily energy requirements. Under these conditions the territory is abandoned.

Some animals only defend territories during the breeding season. In some cases the males defend territories, called *leks, where the females come for mating. In some cases the territory contains vital resources that the females require, and the males gain access to females by controlling these resources. In the American bullfrog (*Rana catesbeiana*), for example, some males achieve much better mating success than others. These are the ones that are able to defend a territory containing the best egg-laying sites.

theory of mind A theory claiming that some animals are capable of **mental state attribution** (i.e. one animal attributes a particular mental state to another). Thus a young chimpanzee (*Pan* sp.), seeing some food not noticed by others, might give an *alarm call to divert the attention of seniors, and then take the opportunity to secretly grab and eat the food. Some theories would assume that this was a matter of ordinary *learning, and that the young chimp had simply learned the alarm-trick to obtain food secretly. A theory of mind would say that the young chimp induces the others to **believe** that there is a predator in the vicinity. A belief is a mental state, and in this case it is attributed by the young chimp to other chimps. In other words, he believes that they believe there is a predator in the vicinity.

Theories of mind imply that the animal is capable of *cognition, and exhibits true *intention. Theories of mind are deployed by some scientists in investigating apparent *deceit, *teaching, and *self-awareness in animals. The evidence favouring theories of mind is controversial.

thermoreceptors *Sensory receptors that are sensitive to temperature. They are sometimes differentiated into heat and cold receptors. Thermoreceptors occur in very many species. The cockroach (*Periplenata* sp.) has thermoreceptors in the legs that are sensitive to ground temperature, and some on the antennae sensitive to air temperature. Many reptiles have a well-developed temperature sense, with thermoreceptors in the brain as well as in the skin. Pit vipers (Crotalidae) have special pits on the face that are sensitive to infra-red radiation and are shaped so as to give a directional temperature sense.

In birds and mammals, thermoreceptors are important in *thermoregulation. Thermoreceptors in the skin give early warning of temperature change, and thermoreceptors in the central nervous system influence warming and cooling mechanisms, such as shivering, panting, feather and fur movements, etc.

thermoregulation A form of *homeostasis by means of which body temperature is controlled. Most animals have an optimum body temperature around which they function most efficiently. Below this temperature the metabolic rate declines progressively, muscular activity diminishes, and the animal may become torpid. Above the optimum temperature, metabolic rate rapidly increases, but there is an upper limit to the temperature at which bodily processes remain viable. For most species this limit is around 47°C.

The metabolic reactions of the body produce heat continuously, and the more active the animal, the greater the rate of heat production. Many animals are cold-blooded, in the sense that their body temperature tends to conform to that of the environment. Such animals are largely dependent upon behavioural thermoregulation to maintain their optimum body temperature. Animals that gain heat primarily from external sources, such as sunlight, are called **exothermic**, while those that gain heat primarily from internal processes are called **endothermic**. Exothermic animals can warm themselves by seeking warm habitats, or exposing themselves to sunlight. Some enhance the effects of sunlight by changing colour, since dark surfaces absorb more radiant heat than do light-coloured surfaces.

The main problem for endothermic animals is to lose heat, because when cold they can easily increase their metabolic activity by physiological means, or by increasing muscular activity, as in shivering. The most common way of losing heat is by evaporation of water, either from the skin, as in sweating, or via the respiratory system, as in panting. The warm-blooded, endothermic animals, mainly birds and mammals, exhibit true thermal homeostasis, maintaining a constant body temperature despite changes in environmental temperature. Their high metabolic rate provides an internal source of heat, and their insulated body surface prevents uncontrolled dissipation of this heat. The brain receives information about the temperature of the body, and is able to exercise control over the mechanisms of warming and cooling.

thinking Mental simulation of possible future outcomes. If thinking requires some *cognition (manipulation of explicit knowledge), rather than ordinary *learning and *memory, then the issue of thinking in animals remains controversial.

In the study of *problem solving by animals, primates tend to learn rules (such as match-to-sample) more quickly than other species, but such evidence of thinking is very indirect. Although a variety of types of task have been studied, particularly in primates, hard evidence of true thinking, as opposed to quick learning, remains elusive.

thirst A state of *motivation that arises primarily as a result of dehydration of the body tissues. All animals require water to maintain their metabolic processes. All animals lose water by a variety of routes, including excretion, *thermoregulation, and evaporation from the body surface. Lost water must be replenished by *drinking, by eating foods containing water, or by absorption through the skin.

Thirst affects behaviour in three main ways: by increasing the *tendency to seek water; by cutting down on food intake; and by altering thermoregulation. The increased tendency to drink can arise as a result of primary or secondary thirst. **Primary thirst** results from tissue dehydration, and this is monitored by the brain, via changes in the constitution of the blood. **Secondary thirst** is purely psychological, and arises as a result of events that are likely to increase dehydration in the future. Many birds and mammals drink when they eat, if water is available. Food that is eaten, unless it has a high water content, is likely to cause dehydration, both directly, and as a result of water lost in eventual excretion. Prandial drinking has the advantage that the water taken with the meal forestalls any dehydration that might arise as a result of food intake. Similarly, many birds and mammals drink in response to rises in environmental temperature. Here the animal is drinking in order to make water available for thermoregulation.

As in other aspects of *homeostasis, thirst is sometimes a victim of compromise in the interests of the total *welfare of the body. In hot weather, in the absence of water, it may be necessary for animals to experience considerable dehydration in the interests of thermoregulation. Similarly, hungry animals may be obliged to run up a water debt, when drinking is not possible.

threat A form of *communication, usually in the form of a *display or *vocalization. Threat displays are often behaviour patterns that have undergone *ritualization during the course of *evolution. They may be derived from *displacement activities, *intention movements, or other types of behaviour typical of motivational *conflict.

The main function of threat is to keep rivals at a distance without undue expenditure of energy or risk of injury. The territorial *song of birds, and the *scent marking of mammals serve this function, without direct confrontation with rivals. Other threat displays, particularly display of weapons, may proceed outright *aggression. For example, male walruses (*Odobenus rosmarus*) display their tusks during encounters with rivals.

Threat postures are sometimes used to repel members of other species. These may be similar to those used against conspecifics, or they may take the form of intimidating *warning displays, evolved as a means of *defensive behaviour. For example, The hawkmoth (*Smerinthus ocellatus*) exposes a pair of eye-like markings on its wings when it is disturbed by a predator. Small birds are deterred by this apparent threat.

time A *sensory system, based upon the animal's *biological clock, that enables the an animal to respond to the passage of time, particularly time of day or time of year. Cyclic changes in the environment are monitored in three main ways: (1) there may be a direct response to various changes in external (exogenous) geophysical stimuli; (2) there may be an internal (endogenous) *rhythm that programs the animal's behaviour in synchrony with the exogenous temporal period, particularly a 24-hour period, or a 365-day period; (3) the synchronization mechanism may be a combination of (1) and (2). In this case, there must be an external agent that entrains the endogenous rhythm to the external environment. This is called the **zeitgeber*. For example, a 24-hour cycle of external temperature, with an amplitude of 0.6°C, is sufficient to act as a *zeitgeber*, and to entrain the daily activity cycles of lizards (*Draco* spp.).

An animal may use many features of the external environment to gain information about the passage of time. The most important of these is the apparent movement of celestial bodies, such as the sun, moon, and stars. Such influences have been much studied in birds and in bees (Apidae). In addition, it is possible that animals can obtain time cues from changes in environmental temperature, barometric pressure, and magnetic phenomena.

Tinbergen, Nikolaas (1907–88) Dutch biologist who, together with Konrad *Lorenz, founded modern ethology. He studied biology at the University of Leiden, In 1938 he visited Konrad Lorenz at Altenberg. During the Second World War he was interned in a hostage camp in The Netherlands. Afterwards he became Professor of Zoology at Leiden. In 1949 he was invited to become Lecturer in Zoology at Oxford, where he founded the Animal Behaviour Research Group. He retired in 1974. In 1973, together with Konrad Lorenz and Karl von *Frisch, he was awarded the Nobel Prize for medicine.

tolerance The ability to tolerate extreme values of environmental factors, such as temperature, humidity, etc. Some species have the ability to change their tolerance range by *acclimatization. For example, the small tree lizard *Urosaurus ornatus* normally has a temperature tolerance range with a maximum of 43.1°C. However, by maintaining these animals in the laboratory for a period of 10 days, at a temperature of 35°C, compared with the normal temperature of 22–26°C, it is possible to raise their maximum temperature tolerance to 44.5°C.

In the natural environment, the distribution of animals frequently reflects their tolerance along environmental gradients of temperature, altitude, etc. Population density will be high only in those areas where tolerance optima overlap. When given a choice, most animals prefer certain (optimal) values of particular environmental variables, although these preferences can change as a result of acclimatization. For example, temperature tolerance in fish is often directly correlated with monthly changes in their habitat temperature. Many fish also show temperature preferences that are related to their state of acclimatization.

tool using The use of an external object as a functional extension of the body, in attaining an immediate *goal. This excludes many cases of manipulation of objects by animals that bear a superficial resemblance to tool using but lack certain elements. For example, there are birds that drop food items onto rocks to smash them open. These birds are not using the rock as an extension of the body, but an Egyptian vulture

(*Neophron percnopterus*) that throws a stone at an egg to break it, is using the stone as a tool. A nest might be regarded as an extension of the body, fashioned for rearing the young, but this is not really a short-term goal, and a nest is not usually regarded as a tool. Manipulation of environmental matter to use as nest material might seem like tool use, but most biologists would distinguish between the material being manipulated and the means by which the material is manipulated. Knitting needles are normally regarded as tools, but not knitting wool.

In some cases an animal may merely pick up an object and use it as a tool, but in some animals effort is put into altering the object to make it more effective as a tool. For example, the Galapagos woodpecker finch (*Cactospiza pallida*) probes for insects in crevices in the bark of trees by means of a cactus spine, or twig, held in the beak. The birds select spines and twigs that are appropriate to the task, and may break them to a convenient length.

Tool using has been most thoroughly studied in primates. It has been observed in the field and studied in the laboratory as part of investigations into *problem solving and *intelligence. It is commonplace for branches to be modified to make them suitable for reaching and probing for food. Leaves may be formed into sponges to obtain water from holes for drinking or washing.

trade-off Balancing of priorities, either as an aspect of *function, or of *motivation. *foraging efficiency is usually a matter of trade-off among competing priorities. These may include energy gained versus energy spent, energy gained versus risk of predation, and energy gained versus losses to rivals.

tradition A form of apparent *imitation as a result of which juveniles copy the behaviour or their elders. Traditional behaviour results from *learning migratory routes, *song, and other aspects of *imprinting, in situations that juveniles normally encounter while living with their parents. For example, geese (Anatidae), ducks (Anatidae), and swans (*Cygnus* sp.) migrate in flocks composed of mixed juveniles and adults. The juveniles learn the route that is characteristic of their population, stopping at traditional rest places, and over-wintering localities. Reindeer (*Rangifer tarandus*) show fidelity to traditional migration routes and calving grounds. Tradition is the inevitable consequence of the circumstances in which the young are raised, and of the tendency of young animals to become imprinted upon their habitat, their parents, and their peers.

trial and error A form of *learning in which a movement or manipulation leads to particular consequences. The animal then forms an association between its behaviour and the consequences, and the latter are said to provide *reinforcement for the learning process. For example, a domestic cat (*Felis catus*) may paw at a latch so that a door swings open. Access through the door may provide positive reinforcement, and the cat will be more likely to paw at the latch on a future occasion. If opening the door had led to an unpleasant experience for the cat, the reinforcement may have been negative, and the cat would be less likely to paw at the latch in the future.

tropotaxis A type of *orientation which requires simultaneous comparison of the intensity of stimulation received at two or more receptors, thus enabling the animal to strike a balance between them. It can then steer a course directly towards, or away from, the source of stimulation. If one of two sensors in blocked or covered, then the animal moves in a circle. When provided with two sources of stimulation, such as two light sources, the animal may take a median course, maintaining a balance between the two sensors.

U

unconditional response and unconditional stimulus Terms introduced by *Pavlov in his studies of *classical conditioning. The animal forms an association between a stimulus that is already significant to it (the unconditional stimulus) and a neutral stimulus (the conditional stimulus). As a result of *learning, the animal's *response to the unconditional stimulus (the unconditional response) comes to be elicited by the conditional stimulus alone (this is now called the conditional response). (Note that the terms unconditioned stimulus and unconditioned response are mistranslations of the original Russian, and are no longer used.)

utility A term used in *animal economics to denote the quantity maximized by the individual animal in the process of rational *decision-making. As in human microeconomics, utility is a notional measure of the psychological value of the consequences of an action (e.g. buying goods). It is notional, because we do now know how utility is incorporated into the decision-making process, only that animals behave as if they were maximizing some quantity, called utility. The evidence that animals behave in this way comes primarily from studies of *rationality and *demand functions in animals. The concept of utility is important in the study of animal *welfare, and utility is sometimes seen as synonymous with welfare.

vacuum activity Activities that occur in the apparent absence of the appropriate external stimuli. For example, canaries (*Serinus canaria*) deprived of nest material will perform the movements of weaving material into a non-existent nest. However, an observer can never be sure that the animal has not noticed some stimulus, which by stimulus *generalization, stands for the missing external stimulation.

vigilance A state of readiness to detect certain specific events occurring unpredictably in the environment. Thus, birds often interrupt their feeding behaviour to look around. They do this more often if they sense danger. Thus the level of vigilance is partly a matter of *motivation. The vigilance of a sexually motivated animal may be directed at potential mates, while that of a mated male may be directed at potential rivals. Thus vigilance has elements of *attention to specific stimuli. Vigilance is costly in terms of time and energy, and social animals can benefit from shared vigilance. Birds feeding in flocks look up less often than birds feeding alone, and they are thus able to save time and energy without endangering themselves, provided that other members of the flock raise the *alarm.

vision The detection of light by eyes and the behavioural responses produced. The eye signals to the *brain, or central nervous system, attributes of the light detected, such as its intensity, spatial distribution, wavelength, and variation over time.

Vision is based upon the detection of electromagnetic radiation. The electromagnetic spectrum covers a wide span, of which the visual spectrum is a very small proportion. The energy content of electromagnetic radiation is inversely proportional to the wavelength. The long wavelengths contain too little energy to activate the photochemical reactions that form the basis of photoreception. The short wavelengths contain so much energy that they damage living tissue. All photobiological responses are confined to a narrow band of the spectrum between these two extremes.

Although photoreceptor cells are scattered over the body on some animals, such as the earthworm (*Lumbricus* sp.), they are usually gathered together in clusters. The most primitive type of eye is made up of clusters of receptors arranged at the bottom of a depression or pit in the skin. Such an eye can give a rough indication of the direction of light, as a result of shadows cast by the sides of the pit.

The *evolution of the eye can be traced amongst living molluscs. The pinhole eye of *Nautilus* spp. (a primitive mollusc) is a development of the pit eye, the rim having grown inward and the photoreceptor layer forming a **retina**. The pinhole eye works exactly like a pinhole camera, in that light from each point reaches a very small area of the retina, so that an inverted image is formed. From the pinhole eye, the eye evolved a protective layer, presumably to exclude dirt. Inside the eye a primitive lens was formed, as in the snail *Helix* spp. The eye of the cuttlefish *Sepia* sp. is very similar to that of vertebrates. The shape of the lens can be changed by ciliary muscles, thus permitting variation in focus. Moreover, there is an iris diaphragm capable of regulating the amount of light reaching the retina. Thus there has been much parallel evolution of the eye, culminating in the vertebrate type of eye.

Although the basic layout of the eye is the same in all vertebrates, there are a number of variations and specializations. There is variation in *colour vision and binocularity. Binocular vision is possible in animals where the two visual fields overlap. The advantage of binocular vision is that it permits better *perceptual organization, particularly in judgement of distance. The disadvantage is that the field of view is narrowed. A wide field of view is important for animals, such as herbivores, that have to maintain *vigilance for approaching predators. The eyes of many nocturnal vertebrates contain light-reflecting bodies, called **tapeta**. These cause the eyes to glow when they are caught in a beam of light. Light that enters the eye is reflected back, thus increasing the probability of absorption by a photoreceptor, and increasing the sensitivity of the eye.

vocalization The production of auditory signals. The production of sound by means of a vocal apparatus occurs only in some vertebrates, but many other animals produce sounds that have an equivalent biological role. Thus the *song of birds is vocal, whereas that of crickets (*Gryllus campestris*) is produced by **stridulation**, involving frictional movement of modified wings.

Sound signals are designed to provide *communication in particular ecological and social circumstances. In the case of bird song, for example, the two fundamental problems of sound transmission are attenuation and distortion. Attenuation occurs as a result of the spherical spreading of sound waves sent from the source. The further the receiver from the source, the fainter the sound will be, because the energy is physically diluted. Sound attenuation can be increased by weather conditions and obstacles to sound transmission. Higher sound frequencies have greater attenuation, because they are more likely to be absorbed by the atmosphere, especially in hot or humid conditions. High frequencies are also more easily scattered by obstacles. Sound travels faster in warm air, and as the air is usually warmer near the ground, the sound waves tend to be bent upwards causing a sound 'shadow'. Low-frequency sounds, especially where the source is near the ground, tend to be prone to greater interference effects (due to cancellation and summation of sound waves) than high-frequency sounds. Thus the highest and lowest frequencies are not generally ideal for communication purposes.

Sound distortion may result from scattering by foliage and reverberation from rocks and tree trunks. Pure tones are less subject to distortion but suffer greater attenuation through interference. However, when coding, rather than transmission, is of prime importance, then frequency coding is less affected by weather factors than is amplitude coding. Thus the vocalization of small birds, which cannot produce low-frequency sound, can minimize interference by being produced from a high position, well clear of obstacles.

Sounds that serve an *advertisement function should be easily locatable, whereas those that serve as *alarm signals should be difficult to locate, so as not to endanger the caller. The ideal alarm call should begin and end gradually, so as not to provide the timing cues used in auditory location. They should be low pitched, so as to make intensity comparisons difficult, and they should cover a narrow range of frequencies, thus making it difficult to use differences in tonal quality as a location cue. The alarm call should not vary in pitch, because this might enable a predator to compare the time and loudness of the sounds reaching the two ears.

voluntary behaviour Behaviour that results from an act of will. Applied to animals, the term has a more physiological connotation. Voluntary behaviour has long been identified with the **somatic nervous system** that controls the skeletal muscles, as opposed to the **autonomic nervous system** that controls the muscles of the heart, glands, etc. Thus humans can move fingers at will, but cannot move the muscles of their intestines at will. Animals subjected to *operant conditioning can learn to move somatic, but not autonomic, muscles to obtain a reward. Therefore, many scientists, regard operant behaviour as the equivalent of voluntary behaviour in humans. However, this does not imply that animals are capable of acts of will, since this is a condition that is impossible to verify.

W

walking *Locomotion on legs, typical of the arthropods and most land vertebrates. The number of legs varies from two to more than a hundred (in some millipedes; Myriapoda). Two-legged animals tend to have large feet, to aid balance, or a large tail as in kangaroos (Macropodidae). During walking, there is always one foot on the ground, but both feet may leave the ground during running.

Walking is the slowest quadrupedal gait, and the most stable, because there are always three feet on the ground. All faster gaits have unstable phases, when there are two, one, or no feet on the ground.

Insects have six legs, which are generally moved in sets of three, to that the insect is always stable. Animals with more than six legs may have a variety of gaits and can readily maintain stability.

warning An aspect of *communication that functions either to alert conspecifics to danger, or to ward off predators. In the former case, the warning may take the form of an *alarm response that takes the form of specific visual, auditory, or olfactory signal.

In the latter case, the warning is a *deimatic display, an evolutionary strategy in which an animal adopts a *display designed to scare off a predator. For example, the caterpillar of the hawkmoth (*Leucorampha* sp.) normally rests upside down beneath a branch or leaf. When disturbed it raises and inflates its head, the ventral surface of which has conspicuous eye-like marks, and the general patterning of which resembles the head of a snake. Many moths and butterflies have eye-spots on their wings, which they reveal suddenly when disturbed, with the possible effect of frightening the predator.

Sometimes a group of noxious species share the same warning signals. Examples are the black and yellow social wasps (*Vespula* sp.), the solitary wasps of the genera *Bembex*, *Eumenes*, and *Crabro*, and the cinnabar moth caterpillar (*Callimorpha jacobaea*). Birds that have learned to avoid cinnabar caterpillars, because of their noxious taste, will not attack wasps, but birds with no experience of wasps or cinnabar caterpillars, will attack wasps. Since predators need to learn one pattern in order to avoid all species in the assemblage, this **Müllerian mimicry** is of benefit to all individuals and no *deception is involved.

welfare The physiological and psychological well-being of animals. Physiological welfare includes freedom from disease and from excessive *hunger, *thirst, and *stress. Successful breeding in captivity is often taken as a sign of good physiological welfare. High productivity of milk, eggs, etc. is also an indicator of good physiological welfare in farm animals, although intensive production methods may reduce the psychological well-being of the animal.

Psychological welfare becomes an issue when there are obvious signs of *suffering as a result of injury, pain, *stress, or *boredom. Psychological welfare is much more difficult to monitor than physiological welfare. Whereas direct physiological measurements of hormones, etc. can be good indicators of physiological distress, there are no direct psychological measures, except in those cases where psychological distress has physiological implications. Moreover, the question of what mental apparatus is required for some types of psychological distress, such as boredom, remains controversial.

Z

zeitgeber The agent that entrains the endogenous *rhythm to the external environment. For example, a 24-hour cycle of external temperature, with an amplitude of 0.6°C, is sufficient to act as a *zeitgeber* (meaning time-giver) to entrain the daily activity cycles of lizards (*Draco* spp.). The *zeitgeber* can be any feature of the external environment that provides information about the passage of time. The most important of these is the apparent movement of celestial bodies, such as the sun, moon, and stars. Such influences have been much studied in birds and in bees (Apidae). In addition, it is possible that animals can obtain time cues from changes in environmental temperature, barometric pressure, and magnetic phenomena.

Further Reading

Beckoff, M. (ed.) (2005). *Encyclopaedia of animal behavior*. Greenwood Press Westport, Connecticut, USA.

McFarland, D. (ed.) (1987). *The Oxford Companion to Animal Behaviour*. Oxford University Press, Oxford, UK.

Shettleworth, S. (2009). *Cognition, Evolution, and Behavior*. Second Edition. Oxford University Press, New York, USA.

McFarland, D. (2008). *Guilty Robots, Happy Dogs: The Question of Alien Minds*. Oxford University Press, Oxford, UK.

Societies

The leading societies for the study of animal behaviour are:

The Animal Behavior Society (ABS) operating in America. http://www.animalbehavior.org/ABS/

The Association for Study of Animal Behaviour (ASAB) operating in Europe. http://www.asab.org/

Index of Animal Names

Index of Latin Names